The Cambridge Exercise Book for the Science Test

CAMBRIDGE
THE ADULT EDUCATION COMPANY

The editor has made every effort to trace the ownership of all copyrighted material, and necessary permissions have been secured in most cases. Should there prove to be any question regarding the use of any material, regret is here expressed for such error. Upon notification of any such oversight, proper acknowledgment will be made in future editions.

MANUFACTURED IN THE UNITED STATES OF AMERICA

ISBN: 0-8428-8710-5

9 8 7 6 5 4 3 2 1

The Cambridge GED Program

The Cambridge GED Program

Writers

Gary Apple
Owen Boyle
Jesse Browner
Phyllis Cohen
Carole Deletiner
Randee Falk
Don Gerstein
Peter Guthrie
Alan Hines
Jeanne James
Lois Kasper
Rachel Kranz
Gloria Levine
Amy Litt
Dennis Mendyk
Rebecca Motil
Susan Muller
Marcia Mungenast
Thomas Repensek
Ada Rogers
Ann Rowe
Richard Rozakis
Elois Scott
Sally Stepanek
Steve Steurer
Carol Stone
Lynn Tiutczenko
Robin Usyak
Kenneth Uva
Shelly Uva
Tom Walz
Willa Wolcott
Patricia Wright-Stover
Karen Wonderman

Executive Editor

Jerry Long

Senior Editor

Timothy Foote

Project Editors

James Fina
Diane Maass

Subject Editors

Jim Bedell
Diane Engel
Randee Falk
Scott Gillam
Rebecca Motil
Thomas Repensek

Art and Design

Brian Crede Associates
Adele Scheff

Contents

Introduction vii

Biology 1
Earth Science 28
Chemistry 43
Physics 51

Half-length Practice Test 62
Performance Analysis Chart 74

Full-length Practice Test 75
Performance Analysis Chart 96

Simulated Test 97
Performance Analysis Chart 120

Answers and Explanations 121

Introduction

This *Exercise Book for the Science Test* can help you prepare for the GED's Science Test. You can use it along with either *The Cambridge Comprehensive Program for the High School Equivalency Examination* or *The Cambridge Program for the Science Test.*

The Three Sections of this Book

This book has three sections: Exercise, Practice, and Simulation. All three sections share a common purpose—to provide you with practice at reading the types of passages and answering the types of questions you will find when you take the GED. The paragraphs that follow explain what the sections of this book are and tell you how to use them to your best advantage.

The Exercise Section

What is the Exercise Section? The Exercise Section allows you to practice one area of science at a time. It is made up of activities grouped according to the four areas of science covered in Unit V of *The Cambridge Comprehensive Program* and Unit II of *The Cambridge Program for the Science Test.* For example, for the chapter on Chemistry in the Cambridge Programs, you will find a corresponding group of exercises in this book that cover Chemistry.

How to Use the Exercise Section. After you complete a chapter in either of the Cambridge Programs, you can practice by using the corresponding exercises in this book. This allows you to practice with skills you have just studied in order to reinforce them. If you want to, however, you can wait until after you've studied all of the chapters, and complete the related exercises to test all of the science areas at once. It would probably be best for you to complete the exercises in this book before you work on the corresponding Practice Items in the textbooks.

The Practice Tests

What Are the Practice Tests? This book has two Practice Tests that are structured like the actual Science Test of the GED. The questions are *not* grouped according to subject. Rather, the subjects vary from item to item, just as on the real test. The first test, the Half-length Practice Test, is made up of 33 items—half as many as make up the actual GED. The second practice is full-length, the same length as the real test. Both tests give you an opportunity to practice taking a test similar to the GED.

How to Use the Practice Tests. You should take the Practice Tests after you have completed the science instruction chapters in a Cambridge Program and the exercises in this book. It would probably be best to take the

Practice Tests in this book after you complete the Practice Section in either textbook. The Half-length Practice Test should take 47½ minutes. The Full-length Practice Test should take 95 minutes, the same amount of time allowed on the actual GED test.

The Simulated Test

What Is the Simulated Test? The Simulated Test is as much like the actual GED test as possible. It has the same number of questions and is just as challenging as the real test. As on the Practice Tests, the area of science varies from item to item. Taking the Simulated Test will help you determine how ready you are to take the real GED test..

How to Use the Simulated Test. You can take the Simulated Test before or after you've taken the Simulated Science Test in one of the Cambridge Programs. Take the Simulated Test under the same conditions as you will have when you take the real Science Test—work without interruption and do not talk to anyone or consult other materials. You should complete the Simulated Test in 95 minutes, the same amount of time you will have on the actual GED.

Scoring Your Work and Using Your Scores

You will find the Answers and Explanations at the back of the book to be a very useful study tool. When you complete a section in this book, compare your answers to the correct answers given. Whether you answer a question correctly or not, you should read through the explanation. Doing this will reinforce your knowledge of science and strengthen your testtaking skills.

After you check your answers to the items on one of the three tests, you can complete the appropriate Performance Analysis Chart. The charts can help you determine which skills and areas you are strongest in, and direct you to the parts of the Cambridge Programs where you can review those areas in which you need additional work.

Biology

Directions: Choose the one best answer to each of the following questions.

Items 1 to 3 refer to the following passage.

A clone is a group of identical cells or organisms that are descended from a single ancestor. All the cells in a clone contain the same genetic material, making a clone an exact copy of its ancestor.

Cloning can occur naturally and is one way some plants and lower animals reproduce. For example, the common houseplant known as the spider plant sends out shoots from which new plants grow. Sponges and jellyfish are animals that are capable of reproducing by themselves.

In the laboratory, scientists use a variety of methods to make clones. One method is to take the genetic material from one cell and place it in another cell. Other methods involve embryos, which are organisms in the early stages of growth and development. In one such method, the nucleus of an embryo body cell is transferred into the egg cell of another whose own nucleus has been removed. Another method of cloning involves splitting a single embryo into two or more parts. Each part then develops into an independent adult.

1. A "test-tube" baby is conceived by uniting the father's sperm with the mother's egg outside their bodies. The resulting embryo then is placed into the mother's uterus, where the baby is nourished until ready to be born. Which of the following is the reason such a baby is not a clone?

   (1) The baby is conceived in a laboratory.
   (2) The baby develops inside the mother's body.
   (3) The baby is conceived by using two people's genetic material.
   (4) Only one child develops from the resulting embryo.
   (5) The baby is not conceived by dividing a single embryo into two parts.

2. Which of the following is not an example of a clone or clones?

   (1) an African violet grown from a leaf taken from another plant
   (2) a mouse produced by transplanting the nucleus of one embryo into the egg cell of another mouse
   (3) a zebra fish that looks exactly like its father
   (4) two day lilies grown by dividing a single day lily into two parts and then planting the sections
   (5) a single-celled algae plant that divides into two new plants

3. Which of the following is an example of cloning that most likely was done with human assistance?

   (1) A single piggyback plant produces clusters of new little plants on the leaves of adult plants.
   (2) All MacIntosh apple trees are the result of one mutated seed and are genetically identical.
   (3) One amoeba splits into two amoebas, which each split again, and so on.
   (4) A herd of genetically identical cattle is produced.
   (5) A piece of a vine falls to the ground and a new plant grows.

Item 4 refers to the following figure.

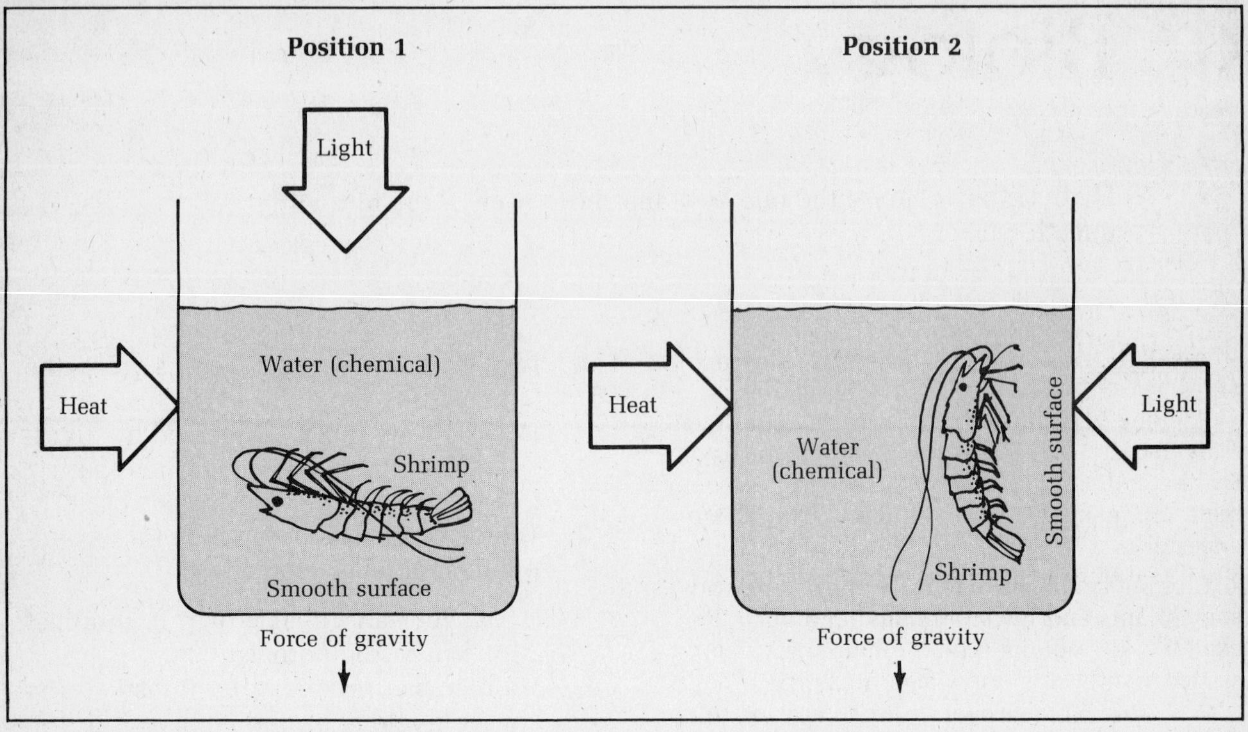

4. Some animals show an unlearned, inborn movement in response to stimuli, such as chemicals, force of gravity, heat, light, or textures. To which of these stimuli is the above shrimp responding?

(1) chemical
(2) force of gravity
(3) heat
(4) light
(5) texture

5. A certain acid is known to reduce the amount of grain produced by a wheat plant while increasing the plant's vegetative growth. The acid, gibberellic acid, is produced by the plants themselves. Which of the following is most likely the reason farmers prefer growing wheat plants that produce low levels of gibberellic acid?

(1) The plants grow taller.
(2) The plants use less energy to make grain.
(3) The plants produce a superior-tasting grain.
(4) The plants have a shorter growing season.
(5) The plants make more grain.

6. A recent scientific study indicated that certain compounds found in plants' reproductive parts can block cancer tumor development. The compounds, found in such foods as beans, rice, and potatoes, prevented the development of cancer during experiments on cultured cells and rodents. In the past, these same compounds had been thought to be antinutritious, because they interfere with the breakdown of proteins by the digestive system.

Which of the following must be assumed in order to conclude that a diet rich in beans, rice, and potatoes can prevent cancer in humans?

A. Both human and rodent cancer cells react to the anti-tumor compounds in a similar way.
B. All cancers are blocked by the anti-tumor compounds.
C. The compounds block tumors when eaten as a part of beans, rice, and potatoes.

(1) A only
(2) B only
(3) A and B only
(4) A and C only
(5) B and C only

Items 7 to 10 refer to the following graph.

The Results of an Experiment Studying the Effects of Light on Plant Growth

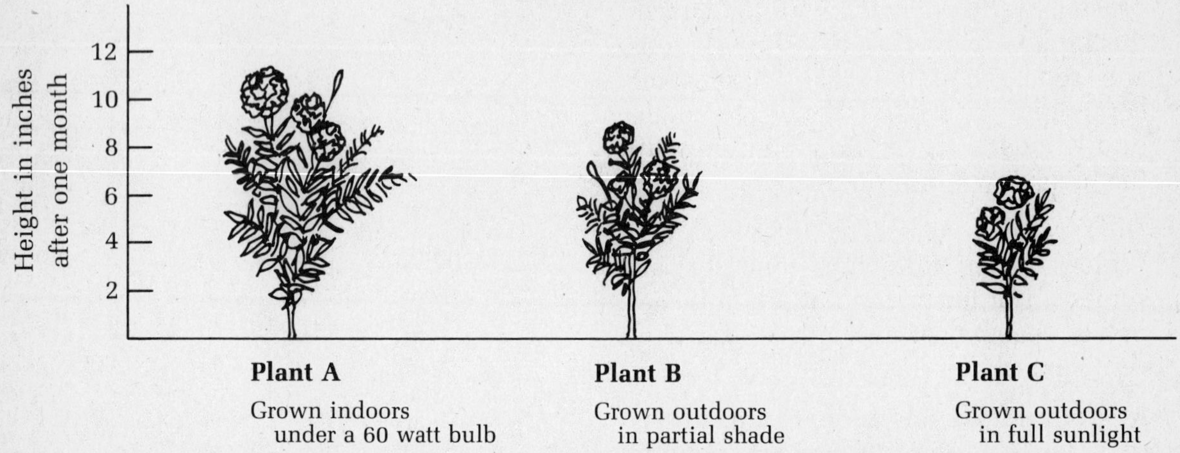

7. Based on the graph, extra light has which of the following effects on marigolds?

(1) More light causes marigolds to produce more seeds.
(2) More light causes marigolds to require more water.
(3) More light causes marigolds to require less water.
(4) More light causes marigolds to grow taller.
(5) More light causes marigolds to grow less tall.

8. You wish to grow the tallest possible marigold. According to the graph, which of the following should you do?

A. Water the marigold frequently.
B. Grow the marigold indoors.
C. Grow the marigold outdoors.

(1) A only
(2) B only
(3) C only
(4) A and B only
(5) A and C only

9. Which of the following statements explain(s) the response of the plants to varying amounts of light?

A. A plant that grows tall under shady conditions requires more energy from sunlight to reach its height.
B. A plant that grows tall under shady conditions increases its chances of reaching enough light for its needs.
C. When a plant has enough light for its needs, it doesn't expend energy on growing tall.

(1) A only
(2) B only
(3) C only
(4) A and B only
(5) B and C only

10. Which of the following factors, if different for each plant, would not cause an error in the experiment?

(1) soil
(2) plant species
(3) light intensity
(4) color
(5) water

Items 11 and 12 refer to the following paragraph.

Trichina worms are parasites that live in the intestines of pigs and humans. They enter humans who eat pork that has not been cooked well enough to kill the worms. In about 16 days, the young trichina worms develop into adults in the intestines and reproduce. The resulting larvae burrow into the muscles; among the muscles most often invaded are those of the eyes, neck, throat, larynx, and tongue. The condition, called trichinosis, causes flu-like symptoms and can damage affected muscles permanently.

11. If the description above is a typical one, which of the following generalizations about the parasite-host relationship is most likely correct?

(1) The relationship benefits both host and parasite.
(2) The relationship harms the host and benefits the parasite.
(3) The relationship kills the host but leaves the parasite unharmed.
(4) The relationship does not harm or benefit the host but benefits the parasite.
(5) The relationship harms both host and parasite.

12. A person goes to the doctor complaining that he has the flu. The doctor's questions and examination reveal that the patient has the following symptoms: diarrhea, vomiting, fever, pain, muscle stiffness, and swelling of the face. The doctor's diagnosis is that the patient likely has trichinosis, not the flu. What factor(s) probably influence his decision?

(1) The patient told the doctor he had eaten pork chops for dinner the previous night.
(2) The vomiting, fever, diarrhea, and pain experienced by the patient would lead only to a diagnosis of trichinosis.
(3) The patient said he recently had been with a person who had trichinosis and probably caught it from her.
(4) The facial swelling indicated that the muscles of the eyes, neck, throat, and tongue were being affected by something other than a flu virus.
(5) The doctor had not treated anyone recently for the flu and it was not the flu season.

---

13. The number of hours of light and darkness in each day will determine when and if a plant flowers. Long-day plants, such as corn, produce flowers when there are 12 to 14 hours of daylight. During which of the following times of the year is a corn plant in North America most likely to flower?

(1) early spring
(2) late spring
(3) summer
(4) early fall
(5) late fall

14. Photoperiodism is the response of a plant to the changing number of hours of light and darkness during the year. Other than flowering, which other plant phenomenon does photoperiodism probably affect?

(1) the loss of leaves in the fall
(2) the production of sugars during the day
(3) the capture of the sun's energy by leaves during the day
(4) the excretion of carbon dioxide from leaves at night
(5) the bending of a plant stem toward the sunlight

Items 15 to 20 are based on the following information.

THE MAJOR LAND BIOMES			
Biome	**Latitude (degrees north or south)**	**Dominant Vegetation**	**Rainfall/year (centimeters)**
tundra	60°–80°	lichens, mosses, grasses and shrubs	10 to 12 cm
taiga	50°–70°	evergreen trees	25 to 75 cm
deciduous forest (trees that lose their leaves in the fall)	10°–55°	oak, maple, hickory, beech, chestnut, and birch trees	75 to 150 cm
grassland	0°–50°	grasses and wildflowers	25 to 75 cm
desert	20°–45°	cacti and other plants specially adapted to conserve water	less than 25 cm
tropical rain forest	0°–20°	more than 100 species of trees; many trees with large, broad leaves; and vines that grow on other plants	200 to 400 cm

Biomes are large areas with a certain type of climax community. On land, a biome is defined by its plant life. The type of vegetation that can develop in an area is limited by its temperatures, rainfall, and other physical conditions. The major land biomes are defined in the table below.

15. Orchids are plants that attach themselves high up on trees. Their roots obtain moisture from the air. Orchids are most likely found in which biome?

(1) tundra
(2) taiga
(3) deciduous forest
(4) grassland
(5) tropical rain forest

16. In one type of biome, nutrients in the soil are quickly washed away by frequent rains. Thus most of the nutrients are found in the living organisms that populate the biome. Which biome is this?

(1) tundra
(2) taiga
(3) deciduous forest
(4) grassland
(5) tropical rain forest

17. Which of the following animals is probably a native of the North American grasslands known as the Great Plains?

(1) alligator
(2) buffalo
(3) monkey
(4) parrot
(5) zebra

18. Which of the following factors most likely accounts for the difference in dominant vegetation between the taiga and the grasslands?

(1) rainfall
(2) temperature
(3) atmospheric conditions
(4) dominant animals
(5) landforms

19. Which of the following adaptations is most likely to help a desert animal survive?

(1) thick fur
(2) ability to obtain water from plants
(3) ability to digest plant fibers, such as cellulose
(4) keen eyes and ears
(5) ability to climb trees

20. Which of the following factors probably explains why more animals live in tropical rain forests than in deserts?

(1) There are more green plants in the rain forest.
(2) Tropical rain forests are not like deserts.
(3) Tropical rain forests are closer to the equator than deserts.
(4) There are more tropical rain forests than there are deserts.
(5) Man can hunt more easily in the open desert and has killed many desert animals.

---

21. For many years, the sewage of Cleveland, Buffalo, Toledo, and Detroit was deposited into Lake Erie. By the 1960s, the lake ecosystem had become "overfertilized." As a result, its plant and bacteria populations increased rapidly. The bacteria took more oxygen from the water than was replaced. Many fish died as the lake's oxygen level dropped below that required for their survival. Scientists feared the lake soon would support no life at all.

The above paragraph illustrates a biological principle. Which of the following choices states this biological principle?

(1) The balance among the parts of an ecosystem is constantly changing.
(2) The balance among the parts of an ecosystem is hard to disrupt.
(3) The relationship among the parts of an ecosystem is hard to determine.
(4) To exist, an ecosystem must maintain a delicate balance among all its parts.
(5) Sewage is harmful to all parts of an ecosystem.

Items 22 to 27 are based on the following information.

Organisms can be classified according to how they obtain nutrients. Below are five types of organisms.

A.	carnivore:	eats animals only
B.	herbivore:	eats plants only
C.	omnivore:	eats plants and animals
D.	producer:	makes own food
E.	saprophyte:	eats the dead or decaying matter of other organisms

Each of the above categories might be used more than once in the following items.

22. In a given day a person eats cereal with milk for breakfast, a tuna sandwich with lettuce and tomato for lunch, and chicken, spinach, rice, and apple pie for dinner. A human being is a(n)

(1) carnivore.
(2) herbivore.
(3) omnivore.
(4) producer.
(5) saprophyte.

23. A mushroom growing on the forest floor obtains nutrients from fallen leaves, rotting logs, and animal droppings. A mushroom is a(n)

(1) carnivore.
(2) herbivore.
(3) omnivore.
(4) producer.
(5) saprophyte.

24. A maple tree converts carbon dioxide and water into carbohydrates during photosynthesis. These carbohydrates are used by the tree for growth. A maple tree is a(n)

(1) carnivore.
(2) herbivore.
(3) omnivore.
(4) producer.
(5) saprophyte.

25. A horse spends its life eating hay and oats given by its owner and grazing in the fields. Occasionally the owner treats the horse to some carrots or sugar cubes. The horse is a(n)

(1) carnivore.
(2) herbivore.
(3) omnivore.
(4) producer.
(5) saprophyte.

26. A lion devotes most of its day to catching antelope, gazelles, and other plant-eaters. A lion is a(n)

(1) carnivore.
(2) herbivore.
(3) omnivore.
(4) producer.
(5) saprophyte.

27. Most spiders weave sticky webs that catch unsuspecting insects as they are walking or flying by. The insect dies in its struggle to get free and is eaten when the spider is ready for it. A spider is a(n)

(1) carnivore.
(2) herbivore.
(3) omnivore.
(4) producer.
(5) saprophyte.

Item 28 refers to the following graph.

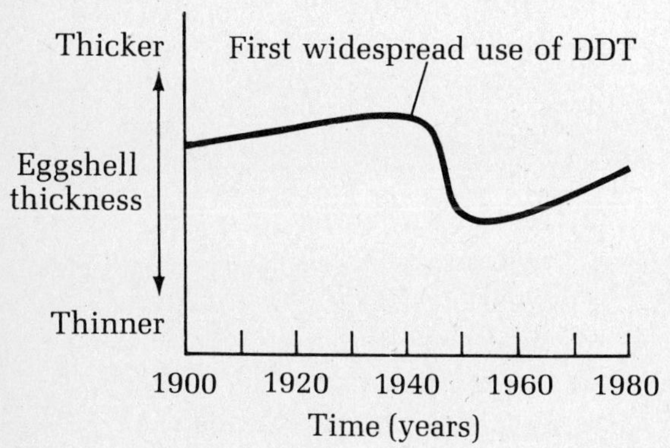

28. Which of the following hypotheses best explains why the widespread use of DDT, an insecticide, nearly destroyed certain species of birds within a few generations? Birds contaminated by DDT

(1) lay fewer eggs
(2) lay eggs with shells too thick for baby birds to peck out of
(3) lay thin-shelled eggs that break under the weight of the parent bird before the baby bird is ready to hatch
(4) eat less and thus lack the nutrients necessary to lay eggs
(5) are unable to mate and thus lay no eggs

29. Monoculture is the farming practice of growing a single crop in a large area. Which of the following facts represent disadvantages of monoculture?

A. The farmer buys fertilizer and insecticide designed for that particular crop.
B. Insect pests that prey on that particular crop can quickly establish themselves throughout the area.
C. The farmer has only one crop to sell.
D. An ecosystem based on one particular crop generally is unstable.

(1) A and B only
(2) B and C only
(3) C and D only
(4) B and D only
(5) A, B, and D only

30. An involuntary response cannot be consciously controlled. Which of the following responses is involuntary for human beings?

(1) breathing slowly when resting
(2) crying when in pain
(3) running from danger
(4) smiling when happy
(5) pupils getting smaller in bright light

Items 31 to 33 refer to the following information.

Heart disease causes more deaths of both men and women than all cancers combined. One of the major risk factors associated with heart disease is high blood cholesterol. Cholesterol is a fatty substance that is a major part of deposits in the blood vessels. These deposits narrow the blood vessels, causing the heart to work harder to pump blood throughout the body.

Scientists have found ways to reduce high blood cholesterol and thus the risk of heart disease. Dietary changes, such as eating less fat and cholesterol, can lower blood cholesterol. A recent study also found that eating grains, such as oats, can help. In addition, a program of regular exercise also reduces blood cholesterol.

31. Which of the following would not help reduce the risk of heart disease?

(1) Eat less cereal.
(2) Drink skim milk instead of whole milk.
(3) Broil or bake meats instead of frying them.
(4) Trim visible fat from meats.
(5) Snack on fruits and vegetables rather than potato chips.

32. The American Heart Association recommends that you eat no more than 300 milligrams of cholesterol a day, about as much as is found in one egg. Which of the following statements is the basis for this recommendation?

(1) Regular exercise reduces blood cholesterol.
(2) A diet low in fat reduces blood cholesterol.
(3) Heart disease is a leading cause of death in the United States.
(4) A diet high in cholesterol leads to increased blood cholesterol.
(5) A diet high in grains reduces blood cholesterol.

33. Which of the following statements is based on fact rather than opinion?

(1) Cholesterol is a major component of blood-vessel deposits.
(2) Athletes never die of heart attacks.
(3) If a food tastes good, it's bad for the heart.
(4) There is no way to avoid heart attacks.
(5) Running is the best exercise for the heart.

Item 34 refers to the following graph.

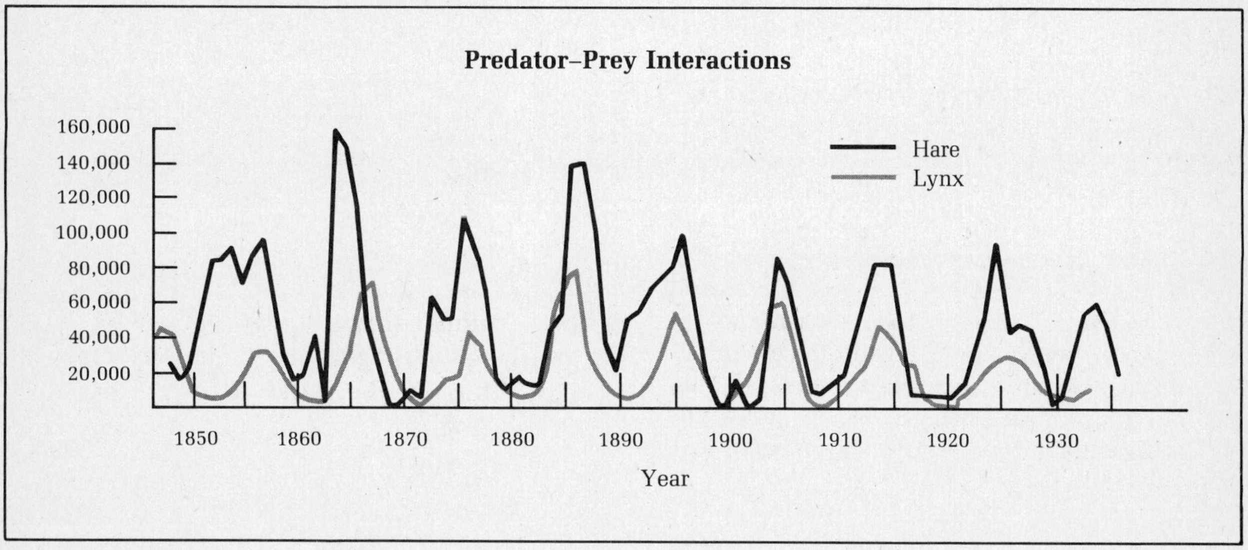

34. The hare is a major foe, or prey, of the lynx. If the above example is a typical one, which of the following is true of all predator-prey relationships?

(1) The peak populations of the predator tend to be larger than the peak populations of the prey.
(2) A decline in the prey population is followed by a decline in the predator population.
(3) Populations of prey and predator peak every 10 years.
(4) Overcrowded conditions are harmful to both prey and predator.
(5) Both prey and predator mark out specific territories, which they defend against others of their species.

35. In a series of experiments, research scientists surgically damaged the inner ear cells in chick embryos and baby chicks. The damage was equal to that caused by a very loud noise, such as that found in some industrial settings or on an airport runway. Within a few days, the researchers discovered damage in areas of the brain that process signals from the air, indicating early hearing loss can affect brain development. If early hearing loss is found to interfere with brain development in humans, which of the following activities or environments should young children avoid?

(1) airplane trips
(2) car rides
(3) factories
(4) playgrounds
(5) movie theaters

Items 36 to 40 refer to the following passage.

Behavior is defined as any external activity or communication by an animal. Behavior includes such activities as taking in nutrition, reproducing, avoiding predators, engaging in social relations, and sleeping. Effective behavior helps either the animal or its offspring and relatives to survive.

Behaviors can be inherited, learned, or both. Simple behaviors tend to be inherited. For example, a reflex is an automatic, inherited response to a stimulus or change in the environment. When a human being is struck on the leg directly beneath the knee cap, the leg jerks. The jerk is an automatic, uncontrolled response to the tap on the leg.

Most animals show learned behaviors as well as inherited behaviors. Learning is the modificatin of behavior by experience. In general, the higher animals depend more on learned behaviors than on those they inherited. Human beings must learn nearly every behavior, including speech, reading, and walking.

Scientists believe most behaviors are an interaction between learning and genetics. Thus, our ability to learn a language is inherited. However, the particular language we learn is based on what we are taught through contact with our environment.

36. Which of the following human behaviors is a reflex?

(1) eating when hungry
(2) shutting the eyes in response to an object coming at the face
(3) drinking when thirsty
(4) recognizing familiar faces
(5) fighting when threatened by another person or animal

37. Which of the following statements is implied in the passage's description and definition of a reflex?

(1) Only mammals have reflexes.
(2) A reflex behavior involves the brain.
(3) A reflex behavior has little or no survival value to the organism.
(4) One reflex response often leads to another reflex response.
(5) A reflex behavior is not learned.

38. Animals often vary their behavior to meet changes in their environment. Which of the following statements best explains the value of this ability to vary behavior?

(1) It increases the animal's reproductive opportunities.
(2) It increases the animal's chances for survival.
(3) It increases the animal's avoidance of predators.
(4) It increases the animal's ability to learn.
(5) It increases the animal's communication abilities.

39. A newly hatched duckling will follow any large object that moves, whether it is a duck, a person, or even a balloon. Once the duckling has made this attachment, the object is forever treated as a parent, no matter how unsuited it might be to the job. This behavior is known as imprinting. Based on this information and that in the passage, which of the following conclusions can be drawn about imprinting?

(1) Imprinting is a learned behavior.
(2) Imprinting is an inherited behavior.
(3) Imprinting is an inherited behavior that is modified by learning.
(4) Birds are the only animals that imprint.
(5) Imprinting is the result of communication between the mother and duckling.

40. Which of the following body parts most likely plays the largest role in how humans behave?

(1) brain
(2) heart
(3) nerves
(4) reproductive organs
(5) spinal cord

Items 41 to 45 are based on the following diagram.

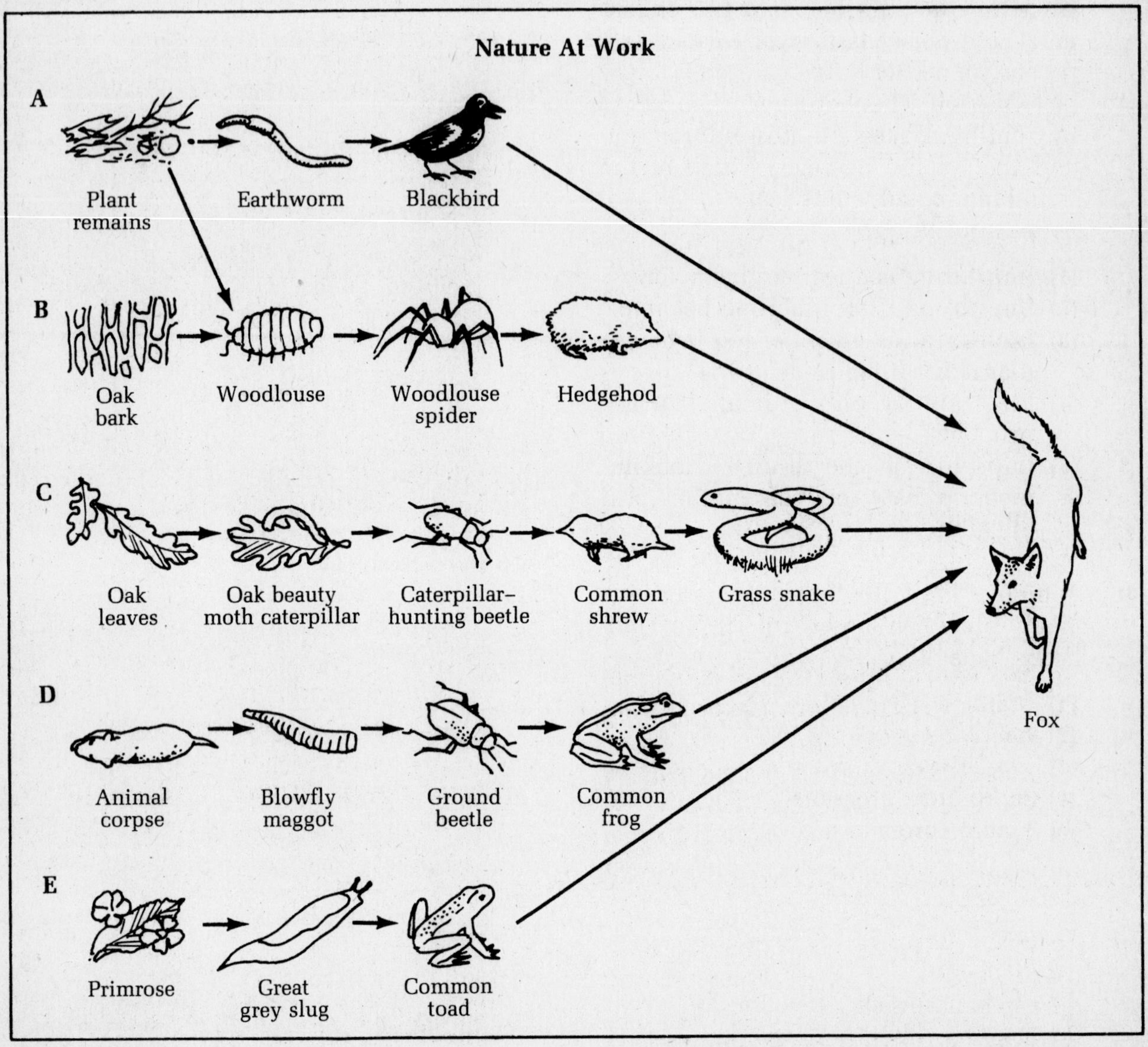

Each consumer in a food chain passes on only a small amount of food energy to the next consumer. Thus each step up a food chain results in the loss of a large amount of food energy.

41. Decomposers are useful in that they break down the remains of plants and dead animals into a form plants can use for nutrition. Which of the following are decomposers?

(1) caterpillar, beetle, spider
(2) moth, maggot, slug
(3) earthworm, woodlouse, maggot
(4) earthworm, caterpillar, maggot
(5) spider, woodlouse, earthworm

42. Why is it rare for a food chain to have more than five links?

(1) Large animals feed at the end of many chains.
(2) Large animals feed at different positions in the same chain.
(3) The animals near the beginning of the food chain become scarce when the chain has too many links.
(4) An animal large enough to be at the end of a very long chain would need more food energy than would be available.
(5) Too many links in a food chain upset an ecosystem's natural balance.

43. Which of the following generalizations is most likely true of all food chains?

(1) An animal feeds only in its particular food chain.
(2) Most animals feed in more than one chain.
(3) Food chains rarely contain more than two links.
(4) There are few connections among different food chains.
(5) Large animals are found only at the end of food chains.

44. Which of the following arrangements of a particular food chain is correct?

(1) grain → person → chicken
(2) chicken → grain → person
(3) person → chicken → grain
(4) person → grain → chicken
(5) grain → chicken → person

45. Which of the food chains in the diagram results in the loss of the most food energy?

(1) A
(2) B
(3) C
(4) D
(5) E

46. Communication through chemicals is widespread among animals. Such chemicals serve as sexual attractants and repellents. They also are used as territorial markers. Some chemicals guide members of a group to a food supply, whereas others serve to identify offspring. Scientists have learned to make a chemical found in the female gypsy moth to lure males to their death. This disrupts gypsy moth reproduction, allowing the population to be controlled. Which of the following categories of chemicals are the scientists probably making use of?

(1) attractant
(2) repellent
(3) territorial marker
(4) food-supply guide
(5) offspring identification

47. A principle of evolution is that organisms adapt to environmental changes in order to survive. An adaptation is an inheritable change that allows a species to adjust to its environment. Which of the following probably would <u>not</u> cause a population of mice to adapt in order to survive?

(1) widespread use of a new pesticide
(2) widespread use of mouse traps
(3) a new strain of virus that is lethal to mice
(4) a change in the usual food supply
(5) a permanent drop in temperature

Item 48 is based on the following table and information.

A scientist was testing the effect of a certain treatment on seeds. During the experiment, the scientist planted bean seeds in window boxes outside a south-facing window. Each seed received two milliliters of water each day. The results are shown in the following table.

	Germinated	**Not Germinated**
Untreated Bean Seeds	50	100
Treated Bean Seeds	80	70

48. What conclusion might the scientist arrive at from this experiment?

(1) The treatment has a positive effect on seed germination.
(2) The treatment has a negative effect on seed germination.
(3) The treatment has no effect on seed germination.
(4) The treatment will work only if the seeds are in a south-facing window.
(5) The treatment will work only if the seeds receive two milliliters of water each day.

Items 49 to 53 refer to the following information.

A cell must be able to perform certain functions necessary to life. Below are five cellular activities.

A. absorption: the passage of materials into the cell
B. digestion: the breakdown of food substances into a usable form
C. reproduction: the production of new cells like the parent cell
D. respiration: the release of chemical energy stored in food
E. synthesis: the chemical combination of simple substances to form more complex substances

Each of the following items describes the job of a particular organelle (a specialized cell part that functions as an organ). For each item, choose the one cellular function the organelle is involved in. Each of the categories above may be used more than once in the following set of items.

49. In the chloroplasts, a plant makes glucose molecules from carbon dioxide and water. The chloroplasts are involved in

(1) absorption.
(2) digestion.
(3) reproduction.
(4) respiration.
(5) synthesis.

50. The lysosomes found in animal cells contain juices that break apart food particles into simpler chemicals that the cell can use. The lysosomes are involved in

(1) absorption.
(2) digestion.
(3) reproduction.
(4) respiration.
(5) synthesis.

51. The cell membrane regulates what substances can come into the cell. The cell membrane is involved in

(1) absorption.
(2) digestion.
(3) reproduction.
(4) respiration.
(5) synthesis.

52. Segments of DNA called genes contain the hereditary information necessary to make a new cell. The genes are involved in

(1) absorption.
(2) digestion.
(3) reproduction.
(4) respiration.
(5) synthesis.

53. The endoplasmic reticulum provides a place for the cell to make lipids or fats. The endoplasmic reticulum is involved in

(1) absorption.
(2) digestion.
(3) reproduction.
(4) respiration.
(5) synthesis.

Items 54 and 55 refer to the following passage.

Recent research by nutritionists indicates that taking a vitamin A supplement can prevent some infections in children. Vitamin A occurs naturally in carrots and leafy green vegetables. The research involved supplementing the diet with extra vitamin A. In one experiment, conducted in Indonesia, scientists found a 70-percent reduction in deaths due to infection. In an Australian experiment, scientists reported a 20-percent decrease in respiratory infections among infection-prone children.

54. Which of the following actions could be part of an experiment to test the theory that extra vitamin A reduces the number of infections a child contracts?

(1) feeding extra carrots to some infection-prone children
(2) testing carrot-eating rabbits for infections
(3) giving extra meat to children with infections
(4) giving a low dose of penicillin daily to infection-prone children and having them eat extra carrots
(5) keeping infection-prone children away from other children as much as possible.

55. Which of the following factors must be assumed in an experiment testing the effect of extra vitamin A on infection-prone children?

(1) The children obtain no vitamin A from other sources.
(2) The children are all the same age.
(3) The children are all the same sex.
(4) The children remain healthy throughout the experiment.
(5) All of the children receive standard medical care throughout the experiment.

Item 56 is based on the following graph.

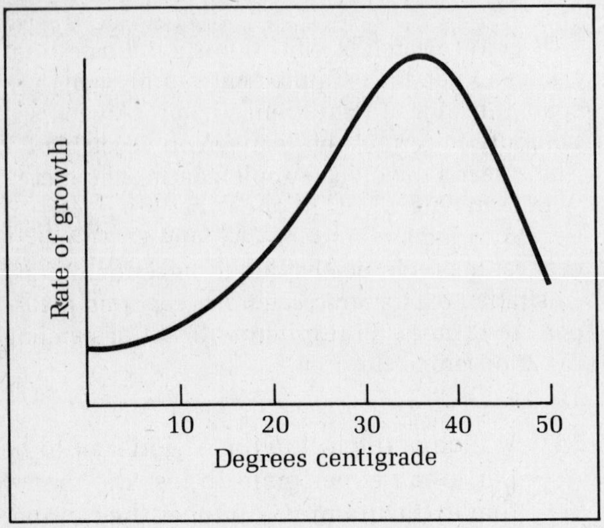

56. For maximum growth, which of the following periods would be the best time to plant a crop? Plant so that the crop is growing when temperatures are between

(1) 0°–10°C.
(2) 10°–20°C.
(3) 20°–30°C.
(4) 30°–40°C.
(5) 40°–50°C.

Items 57 to 60 refer to the following passage.

About one third of the world's rubber today comes from rubber trees. The other two thirds is synthetic, made from petroleum.

Synthetic rubber has many uses, but natural rubber has several advantages over synthetic rubber. Unlike the production of synthetic rubber, producing natural rubber doesn't cause pollution. Also, natural rubber sources are renewable, whereas the petroleum needed to make synthetic rubber is a nonrenewable resource. Furthermore, natural rubber has a special chemical makeup that makes it stronger and longer-lasting than synthetic rubber.

Currently, the need for natural rubber is being met by rubber trees. However, this resource already is being used almost to its limit. Some biologists have suggested turning to the guayule shrub. Found in the southwestern United States and Mexico, the guayule produces rubber that, after processing, is exactly like the product of the rubber tree. In fact, before the invention of synthetic rubber, about half of the rubber used in the United States came from the guayule shrub.

57. Which of the following is the best definition of a nonrenewable natural resource?

(1) one that is nonliving
(2) one that cannot be easily replaced
(3) one that is scarce
(4) one that is chemically produced
(5) one that is used to make synthetic rubber

58. Which of the following supports a race car driver's preference for natural rubber tires over those made of synthetic rubber?

(1) Natural rubber is cheaper than synthetic rubber.
(2) Natural rubber sources are renewable; synthetic rubber sources are not.
(3) Natural rubber is stronger and longer-lasting than synthetic rubber.
(4) Producing natural rubber does not cause pollution; synthetic rubber does.
(5) Natural rubber can be produced from sources within the United States.

59. Which of the following statements gives the main reason a group concerned about the natural environment would favor guayule rubber production?

(1) Guayule rubber production would create jobs in the Southwest.
(2) Guayule cultivation would reduce U.S. dependence on foreign rubber.
(3) About one third of the world's rubber comes from rubber trees.
(4) Natural rubber is strong and long-lasting.
(5) Natural rubber production doesn't cause pollution.

60. Which of the following aspects concerning guayule rubber production would most concern a Southwestern farmer?

(1) The guayule is a desert shrub that grows well in the Southwest.
(2) The United States uses one fifth of the world's rubber.
(3) Guayule rubber processing would provide employment to people in the Southwest.
(4) Guayule rubber is as strong as rubber from rubber trees.
(5) Guayule rubber lasts as long as rubber from rubber trees.

61. Scientists believe that life on Earth began when simple molecules, such as amino acids, came together to make complex molecules. These complex molecules were able to reproduce themselves. Exactly how this happened is not known. However, a variety of models have been tested experimentally. In one such model, the amino acids attached to the surfaces of minerals, which served as patterns for the combination of the simple molecules.

Which of the following statements is an unstated assumption in the theory of life's origin on Earth, as described above?

(1) Amino acids cannot reproduce themselves.
(2) Complex molecules are made up of amino acids.
(3) Amino acids can attach to mineral surfaces.
(4) Complex molecules are alive.
(5) Simple molecules existed before complex molecules.

Item 62 is based on the following figures.

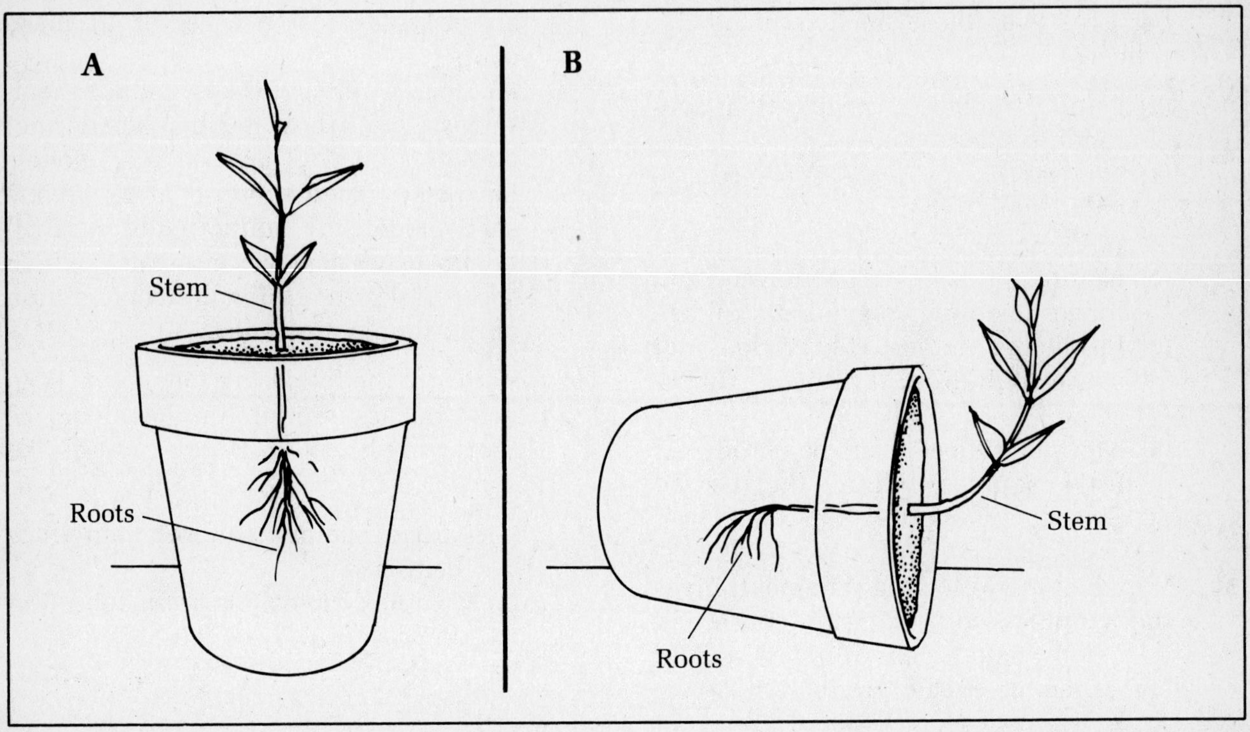

62. A flower seed is planted on the side of a steep hill. Which of the following descriptions gives the position of the resulting stem and roots?

(1) The stem and roots will grow upward.
(2) The stem and roots will grow downward.
(3) The stem and roots will resemble those in Figure A.
(4) The stem and roots will resemble those in Figure B.
(5) The stem and roots will grow straight out from the hill so that they appear tilted to an observer.

63. A long-term study of Harvard-educated men found that the men who walked nine or more miles a week were less likely to die young than those who walked three miles or less. Those who practiced light sports a few hours a week lived longer than those who had no exercise at all. What conclusion can be made as to why some men in the study lived longer than others?

(1) Some men walked nine or more miles a week and some walked three.
(2) Regular, moderate exercise increased the life span of some of the men.
(3) Men who practiced sports activities lived longer than those who did not.
(4) All the men were Harvard-educated, but they had different backgrounds.
(5) Men who had no exercise at all died youngest.

64. After a tree was transplanted from one area of a park to another, its leaves shriveled and the tree died. What best accounts for the tree's death?

(1) disease
(2) rapid water loss because of hot weather
(3) root damage
(4) insect damage
(5) old age

Items 65 to 68 refer to the following passage.

In any environment, each species of plant or animal occupies its own niche. That is, each species has its own particular way of obtaining nutrients and its own place to live and reproduce. Each species in turn has certain effects on its environment. It produces particular wastes and by-products. For example, a plant gives off oxygen as a by-product of photosynthesis.

Although different species of animals live in the same area, they tend to occupy different niches. For example, robins eat mainly earthworms and insects, whereas the golden-shafted flicker bird eats mostly ants.

Each species occupies its niche for the whole of its evolutionary life, which averages about a million years. During its early development, a new species is found in just a few areas. If conditions are right, the species later spreads out in ever-widening circles around its home territory, until it reaches conditions to which it is not adapted. Spreading might be blocked by biological barriers, such as competing species, or by physical barriers, such as water or mountains.

65. Which of the following organisms would not occupy a niche in or on a typical lawn?

(1) decomposer of dead or waste matter
(2) insect eater
(3) grass eater
(4) fish eater
(5) producer of food energy

66. Which of the following is most likely the main barrier to the further spread of the human species at this time?

(1) lack of land on which to expand
(2) competition from other meat eaters
(3) competition from other plant eaters
(4) mountains
(5) deserts

67. The earliest humanlike remains have been found on the continent of Africa. This is evidence for which of the following theories?

(1) Africa today is the continent with the best environment for human beings.
(2) Africa is home to the widest variety of species.
(3) Human beings evolved in Africa about a million years ago.
(4) No other organism in Africa occupies the human species' niche.
(5) Africa is the place where the human species first evolved.

68. If a species is known to have existed for about one million years, what most likely will happen in the future for that species?

(1) It will thrive indefinitely because it has proven it can survive a long time.
(2) It probably will cease to exist in its present form.
(3) It will have to occupy a new niche in order to survive.
(4) It will spread to areas other than those it previously has inhabited.
(5) It will mate with a newer species to ensure survival.

Item 69 refers to the following figure.

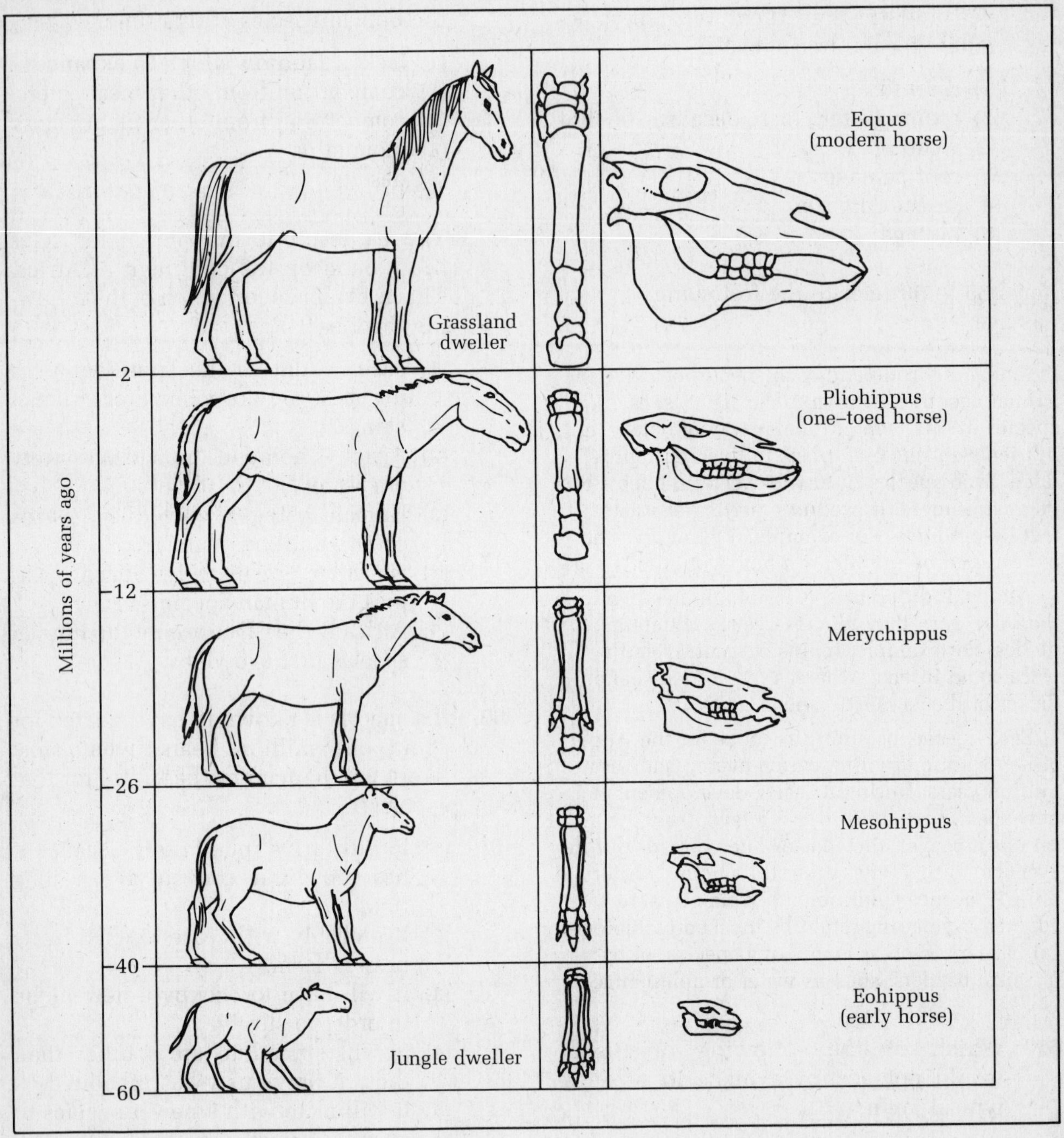

69. Which of the following is a hypothesis that could be used to explain the fact that the species Equus closely resembles Eohippus?

(1) Equus evolved from Eohippus.
(2) Equus and Eohippus have a similar facial structure.
(3) Eohippus lived millions of years before Equus walked the Earth.
(4) Eohippus and Mesohippus both had an extended middle toe that could have developed into Equus' hoof.
(5) Equus and Eohippus occupy the same ecological niche.

70. Social groupings are common among two types of animals—insects and vertebrates, or animals with backbones. Social groups have several adaptive advantages. Animals who live in these groups often find it easier to obtain food and find mates. A group of animals also is better able to defend and protect individual members. Which of the following organisms would you expect to find defending a member of its social group?

(1) bacterium
(2) cactus
(3) clam
(4) bee
(5) worm

71. Protozoa are the most animal-like organisms of the protist kingdom. Although they are one-celled, they do share some important traits with members of the animal kingdom. Most are able to move about freely on their own; they have a whiplike structure that they use to move through their environment. Unlike plants, protozoa do not make their own food. Instead, they must obtain and digest nutrients from the outside environment.

Based on this information, which of the following statements is most likely the reason protozoa are <u>not</u> classified as part of the animal kingdom?

(1) They are one-celled organisms.
(2) They can move around on their own.
(3) They do not make their own nutrients.
(4) They move with the aid of a whiplike structure.
(5) They must digest food into a usable form.

Items 72 to 74 are based on the following table.

SELECTED VITAMINS		
Fat-Soluble	**Function**	**Good Sources**
Vitamin A	Healthy skin and eyes; growth	Spinach, carrots, apricots, fish-liver oil, liver
Vitamin D	Absorption of calcium and formation of healthy bones	Sunshine, eggs, fortified milk, fish-liver oil
Vitamin K	Blood clotting	Green and yellow vegetables; also synthesized by intestinal bacteria
Water-Soluble		
Vitamin C (*ascorbic acid*) (uh SKAWR-bik-AS-uhd)	Healthy connective tissues; helps in iron absorption	Citrus fruits, broccoli, spinach, strawberries, potatoes, cabbage
Vitamin B_1 (*thiamine*) (THIY-uh-muhn)	Release of energy from carbohydrates	Whole-grain products, peas and other legumes, pork

72. A woman is found to have iron-deficiency anemia. Her doctor advises her to eat foods that increase iron absorption along with foods rich in iron. Which of the following fruits would help aid iron absorption?

(1) apricots
(2) bananas
(3) lemons
(4) peaches
(5) plums

73. Rickets is a vitamin-deficiency disease that leads to soft bones. Before its cause was discovered, it was found mainly in children who lived in northern cities that received little sunlight. Rickets is caused by a deficiency of which vitamin?

(1) vitamin A
(2) vitamin B_1
(3) vitamin C
(4) vitamin D
(5) vitamin K

74. Which of the following statements provides the best evidence that lack of vitamin C causes the disease scurvy?

(1) Children with scurvy do not grow well.
(2) People with scurvy do not heal well.
(3) Scurvy in adults can cause bleeding gums and loose teeth.
(4) People with scurvy recover when citrus fruits are added to their diet.
(5) Extreme cases of scurvy cause the breakdown of connective tissue.

75. In the early 1600s, Jean-Baptiste van Helmont planted a tree in a pot of soil after weighing the tree and soil separately. For five years, the only substance van Helmont added was water. The tree was kept in open air. At the end of this time, he again weighed the tree and soil. The tree had grown and gained much weight, but the soil still weighed about the same amount. Van Helmont concluded that the extra weight of the tree came from water alone. But in the late 1800s, Joseph Priestly proved that plants take in carbon dioxide as well as water. What was the flaw in van Helmont's reasoning?

(1) He did not wait long enough before drawing his conclusion.
(2) He did not keep track of the amount of water he gave the tree.
(3) He did not measure exactly how much the tree had grown.
(4) He did not take into account that the tree was exposed to water and air.
(5) He did not take into account the nutrients the tree got from the soil.

Items 76 to 79 refer to the following passage.

The current rapid population growth of human beings has had its disadvantages. The more people there are, the more we affect the ecosystems around us. Highly developed countries like the United States cause environmental problems, such as air and water pollution. The pollution harms many living things. Highly developed countries also use a lot of the world's natural resources, some of which cannot be replaced for many thousands of years. Fossil fuels, such as coal and oil, are burned for energy, giving off poisonous gases as a by-product. Water is used in industry and agriculture. Trees are cut down to be used as fuel and building material.

People in less-developed countries also destroy ecosystems. They harvest forests for wood and clear them to make more living space. All over the world people destroy ecosystems whenever they plow up land to make farms or cities.

Biologists worry about the destruction of ecosystems. When an ecosystem disappears, the living things within it must find a new place to live. Some species cannot adapt and become extinct, never again to be seen on Earth. Biologists believe that every species is valuable for what we can learn from it. In addition, many medicines were found first in natural substances. Plants that face extinction today might contain substances that can fight cancer or other diseases.

76. According to the passage, which of the following is the ultimate cause of the destruction of many of the world's ecosystems?

(1) air pollution
(2) rapid human population growth
(3) burning of fossil fuels
(4) use of water by agriculture and industry
(5) extinction of certain species

77. Which of the following could explain the fact that the United States has an air pollution problem?

(1) The United States has a large population.
(2) The air above the U.S. is sensitive to pollution.
(3) The wind carries pollution to the U.S. from around the world.
(4) Many trees are burned for energy in the U.S.
(5) The U.S. burns a great deal of fossil fuels for energy.

78. Which of the following measures would help preserve the environment?

A. Limit the use of fossil fuels.
B. Install filters in smokestacks.
C. Recycle more glass and paper products.
D. Increase the number of productive farms.

(1) A and B only
(2) B and C only
(3) A, B, and C only
(4) B and D only
(5) C and D only

79. Which of the following statements best supports the biologists' argument that every species is valuable for what we can learn from it?

(1) Penicillin was developed from a mold that grows on bread.
(2) Gasoline is derived from oil.
(3) Trees provide oxygen for animals.
(4) Animals and plants are dependent on bacteria to break down wastes.
(5) Coal is formed from trees' remains over millions of years.

Item 80 refers to the following information and table.

Below are the results of a stress test using two groups of human beings. The test subjected one group to unusual mental or physical stress. The white blood cells of both groups were counted before and after the test. White blood cells are an indicator of the body's ability to fight off infection. A decrease in white blood cells might make a person more susceptible to disease.

RESULTS OF A STRESS EXPERIMENT

	White Blood Cell Count
Unusual mental stress	Decreased
Unusual physical stress	Decreased
Control (no stress)	The same

80. Based on the experiment's results, which of the following life situations might make a person more susceptible to disease?

A. starting a new job
B. the death of a loved one
C. walking around the block
D. reading the newspaper

(1) A only
(2) B only
(3) C only
(4) A and B only
(5) C and D only

Check your answers on pages 121–124.

Earth Science

Directions: Choose the one best answer to each of the following questions.

Items 1 and 2 refer to the following passage.

The rising of deep ocean waters to the ocean surface is called an upwelling. Upwellings occur when wind blowing across the surface of the ocean pushes warm surface water out to sea. Cold water from below the surface rises to replace the surface water. Upwellings are common occurrences along coastlines, particularly along the west coasts of the United States and South America.

Upwellings are important because plants and animals that live near the surface of the ocean are constantly consuming essential nutrients. The rising of deep ocean waters replaces these nutrients. Upwellings are important also because they moderate the temperature of ocean water by bringing deep waters to the surface.

1. Which of the following factors interact to cause upwellings?

(1) water and temperature
(2) wind and water
(3) cold water and warm water
(4) plants and animals
(5) plants and nutrients

2. Every few years, upwellings fail to occur as expected. Which of the following situations would be a likely consequence of such an event?

(1) The fishing industry of Peru experiences record-breaking catches.
(2) Fish die in record numbers, but other sea organisms thrive and multiply.
(3) Ocean water becomes so cold that people cannot swim, even in the warmest weather.
(4) A "heat wave in the sea" causes many fish and other ocean creatures to die.
(5) Fishers along the California coast find that the fish in their catches are larger and healthier than usual.

Items 3 to 6 refer to the following passage.

The earth's primary source of heat is the sun. The sun's energy, in the form of ultraviolet radiation, is absorbed by the Earth and changed into heat. Later the Earth radiates this energy back into space in the form of infrared radiation.

Infrared radiation, unlike ultraviolet radiation, cannot penetrate the Earth's atmosphere. Carbon dioxide and other gases in the atmosphere absorb the infrared rays, forming a kind of insulation around the Earth. This phenomenon is known as the "greenhouse effect" because the carbon dioxide and other gases act like glass in a greenhouse, keeping heat from escaping.

Because most of the infrared radiation from the Earth is absorbed by carbon dioxide, the amount of this gas in the atmosphere is very important. Too little carbon dioxide would allow too much infrared radiation to escape into outer space. Too much carbon dioxide would cause abnormally large amounts of infrared radiation to be absorbed. Increasing amounts of carbon dioxide would cause the greenhouse effect to increase and the temperature of the Earth's atmosphere to become hotter and hotter.

Many scientists are concerned about the amount of carbon dioxide being released into the atmosphere by the burning of fossil fuels, such as coal and oil. They believe that, over an extended period of time, this increase in carbon dioxide could cause the Earth's temperature to rise dramatically. They point out that the consequences of such a temperature change could be disastrous, as farmland is turned into desert and the earth is flooded by the melting polar icecaps.

3. Which of the following is an accurate description of the greenhouse effect?

(1) a method for using glass as a form of insulation
(2) a plan that many scientists believe would reduce air pollution
(3) a way gases in the Earth's atmosphere affect the Earth's temperature
(4) the way ultraviolet radiation is absorbed by the Earth
(5) an effective method for growing plants in cold temperatures

4. Compared to the Earth's atmosphere, the atmosphere surrounding the planet Venus contains much more carbon dioxide, whereas the atmosphere surrounding the planet Mars contains much less carbon dioxide.

Based on the information provided, which of the following is a reasonable conclusion?

(1) The greenhouse effect does not exist on Venus and Mars.
(2) The greenhouse effect exists on Venus, but not on Mars.
(3) Temperatures on Venus are much colder than on Earth or on Mars.
(4) Temperatures on Venus are much warmer than on Earth or on Mars.
(5) Ultraviolet radiation cannot reach Venus as easily as it can reach Earth and Mars.

5. Which of the following statements represents a fact rather than an opinion?

(1) The burning of coal by industry increases the amount of carbon dioxide in the atmosphere.
(2) By the year 3000, the Earth's average temperature will be 20° warmer than it is now.
(3) An increase in the amount of carbon dioxide in the atmosphere will be disastrous to life on Earth.
(4) Industry should limit the use of fossil fuels until scientists determine the harmful effects of increased carbon dioxide in the atmosphere.
(5) A warmer climate would benefit the Earth as long as the polar icecaps did not melt.

6. A scientist in a laboratory must distinguish between a source of ultraviolet radiation and a source of infrared radiation. Based on the passage, which of the following procedures would prove effective?

(1) Project each type of radiation onto a blank screen.
(2) Determine which type of radiation can pass through a piece of glass and heat a pan of water.
(3) Determine which type of radiation can melt a block of ice.
(4) Pass each type of radiation through a piece of glass, then determine which can melt a block of ice.
(5) Determine which type of radiation can pass through a container of carbon dioxide and heat a pan of water.

7. Cool air cannot hold as much water vapor as warm air. As a result, the cooling of warm, moist air often produces dew or fog. The temperature at which the air becomes saturated with water and condenses to form dew is called the dew point. When the dew point is below freezing, frost will form. Which of the following statements explains why fog present in the early morning often disappears by afternoon?

(1) Fog forms more easily in darkness than in daylight.
(2) Humidity tends to be higher in the morning than in the afternoon.
(3) Warmer temperatures in the afternoon cause moisture to evaporate and return to the atmosphere.
(4) The dew point is higher in the morning than in the afternoon.
(5) Frost that forms in the early morning often turns into fog, which then evaporates when the temperature rises above freezing.

Items 8 to 11 refer to the following article.

The state of Connecticut is sitting on a gold mine. According to the United States Geological Survey, rock samples taken from an abandoned cobalt mine in Meshomasic State Forest indicate a gold concentration of 1.5 to 6 ounces of gold per ton of rock. This concentration of gold is 100 times the amount usually found in North America. Gold mines in South Africa, a major world supplier, produced only 0.18 ounce of gold per ton of rock in 1985.

The site of the gold deposit, which is located about 20 miles southeast of Hartford, was discovered accidentally by a group of geologists and students from the University of Connecticut. Although the discovery was made nearly two years ago, nothing has been done about mining the gold.

"We have only limited amounts of protected space for parkland in this state," explained a Connecticut official. "This particular state land is as precious to us as the gold underneath."

The state forest offers many opportunities for fishing, hiking, and winter sports. Private land adjacent to the forest, which also lies above some of the gold, is in a residential zone that is off-limits to industry—and therefore mining of the gold on private property also seems unlikely.

8. According to the article, what makes the gold deposit in Connecticut so unusual?

(1) It is located in a state forest.
(2) Gold usually is not found in the same place as cobalt.
(3) The gold is highly concentrated in the rock.
(4) Gold usually is not found near a city.
(5) It was discovered by accident.

9. Which of the following aspects of the above situation would be most important to an environmental scientist?

(1) the size of the gold deposit relative to gold deposits in South Africa
(2) the effect that mining the gold would have on Connecticut's economy
(3) the geological factors that might have caused a gold deposit to form in this region
(4) the number of people who use the recreational facilities in Meshomasic State Forest
(5) the effect that mining the gold would have on the land and wildlife in this region

10. A Connecticut industrialist who is in favor of mining the gold deposit claims that Connecticut could become the world's major supplier of gold. What piece of information not included in the article would be needed to support this person's claim?

(1) the amount of gold present per ton of rock
(2) the exact location of the gold deposit
(3) the cost of mining gold
(4) the amount of gold produced by the world's leading gold mines
(5) the extensiveness of the gold deposit

11. A team of geologists discovers an underground area of rock that contains silver. Which of the following methods should they use to determine the concentration of the silver deposit?

(1) Remove several rock samples, then determine the amount of silver by weight compared to the total weight of the rock.
(2) Remove several rock samples, then compare them with rock samples from other silver deposits.
(3) Measure the total area of rock, then estimate the amount of silver per cubic meter.
(4) Contact the United States Geological Survey to find out if other silver deposits exist in nearby areas.
(5) Remove several rock samples, then perform chemical analyses on the silver to determine its quality.

Item 12 refers to the following information.

Air is a mixture of gases. The two most abundant gases in air are oxygen and nitrogen. Oxygen is the gas that supports combustion, or burning. When combustion takes place, oxygen gas is consumed.

12. A scientist wishes to determine the percent by volume of oxygen in air. Which of the following experiments should she carry out to do this?

(1) Burn a candle until it goes out in a fixed volume of air, then determine the change in volume of the candle.
(2) Burn a candle until it goes out in a fixed volume of air, then determine the change in volume of the air.
(3) Burn a candle until it goes out in a fixed volume of air, then weigh the candle to determine its change in mass.
(4) Burn a candle until it goes out in a fixed volume of air, then determine the change in mass of the air.
(5) Burn a candle until it goes out in an open container, then analyze the composition of the candle.

Item 13 refers to the following information.

When an earthquake occurs, shock waves are produced. These waves travel outward from the center, or focus, of the earthquake. The types of waves produced by an earthquake are referred to as P, S, and L waves. P waves travel the fastest and can pass through solids, liquids, and gases. S waves travel less fast than P waves and can pass only through solids. L waves are the slowest earthquake waves. L waves move along the surface of the earth and are responsible for the extensive damage that often accompanies an earthquake.

13. An earth scientist believes that a part of the inner earth consists of hot, liquid rock. Another earth scientist claims that this same portion of the earth consists of solid rock. Which of the following investigations would help the scientists settle the dispute?

(1) Determine the temperature of rock samples from this portion of the earth.
(2) Determine how many earthquakes have occurred in this region.
(3) During an earthquake, follow the course of P waves through this portion of the earth.
(4) During an earthquake, follow the course of S waves through this portion of the earth.
(5) Estimate the amount of damage caused by L waves in this portion of the earth.

14. The measure of a star's brightness is called its magnitude. A star's brightness as it appears from Earth is its apparent magnitude. Astronomers also can determine a star's absolute magnitude. The absolute magnitude of a star is the actual amount of light it gives off. Stars that have the same absolute magnitudes often have different apparent magnitudes because they are different distances from the Earth.

Which of the following situations is most similar to the relationship between a star's apparent magnitude and its absolute magnitude?

(1) A person standing in the foreground of a photograph will appear larger than a person standing in the background, even though both persons are the same size.
(2) The headlights of a car traveling straight ahead appear brighter than the headlights of a car turning a corner.
(3) To persons riding in an airplane, objects on the ground appear much smaller than they really are.
(4) A light signal from an airport control tower will appear brighter on a clear night than on a foggy night.
(5) The sound of a train whistle becomes higher in pitch as the train approaches, then lower in pitch as the train moves away.

Items 15 and 16 refer to the following information and chart.

One of the properties by which a mineral can be identified is hardness. Minerals are classified according to hardness by the Mohs Hardness scale. On the Mohs scale, a mineral is assigned a number from 1 to 10, with 10 being the hardest mineral and 1 being the softest mineral. A mineral can scratch any mineral that is lower in number than itself. A mineral can be scratched by any mineral that is higher in number than itself. If two minerals do not scratch each other, they have the same hardness.

MOHS HARDNESS SCALE

Mineral	Hardness
Talc	1
Gypsum	2
Calcite	3
Fluorite	4
Apatite	5
Feldspar	6
Quartz	7
Topaz	8
Corundum	9
Diamond	10

15. One of the minerals listed on the Mohs scale can scratch calcite and apatite, but not topaz. It can be scratched by quartz and corundum. The mineral is

(1) talc.
(2) gypsum.
(3) fluorite.
(4) feldspar.
(5) diamond.

16. A person suspects that a diamond ring he has purchased is a fake. What could he do to prove his point to the jeweler who sold it to him?

(1) Use the stone in the ring to scratch a topaz stone.
(2) Show that the stone in the ring can scratch a fingernail.
(3) Show that the stone in the ring can be scratched by a diamond.
(4) Use the stone in the ring to scratch every mineral on the Mohs scale.
(5) Show that the stone in the ring can scratch a diamond.

17. A light year is the distance that light travels in 1 year. One light year is equal to approximately 9.5 trillion kilometers. Which of the following statements explains why scientists know that Star X is older than the Earth?

(1) The light from Star X is not as bright as the light from younger stars.
(2) The Earth is believed to be 4.6 billion years old, and the distance between the Earth and Star X is 6 billion light years.
(3) Star X is too far from the Earth to be part of our solar system.
(4) The light from Star X travels faster than the light from younger stars.
(5) Satellite photographs taken of Star X show certain features that indicate its age.

Items 18 and 19 refer to the following information and chart.

The chart below shows the relative humidity and average high temperature for cities around the world on an August afternoon. On the comfort index shown, 1 is the most comfortable and 5 is the most uncomfortable. When the comfort index is 3, half the people feel comfortable.

	Relative Humidity	Average High	Comfort Index
Alice Springs, Australia	25	73	1
Spokane, Wash.	25	83	
Tokyo	66	72	
Cheyenne, Wyo.	39	79	2
Paris	61	75	
London	62	71	
Brussels	69	72	
Athens	34	92	3
Madrid	35	85	
El Paso	36	91	
Rome	43	86	
New York	60	80	
Las Vegas, N.M.	15	102	4
Phoenix	27	101	
Marrakesh, Morocco	37	100	
Aswan, Egypt	37	106	5
Sena Madureira, Brazil (Amazon Valley)	99	93	

18. According to the information given, the comfort index is based on

(1) how many people feel comfortable on a given day.
(2) the amount of relative humidity.
(3) how many cities in the world experience the same weather conditions.
(4) a combination of temperature and relative humidity.
(5) a combination of temperature and the number of people who feel comfortable.

19. Which of the following statements explains why more people feel comfortable in Athens, Greece, than in Sena Madureira, Brazil, even though the temperatures in both cities are about the same?

(1) Sena Madureira is closer to the equator than Athens.
(2) The humidity in Sena Madureira is nearly triple the humidity in Athens.
(3) Athens receives less direct sunlight than Sena Madureira.
(4) Cool breezes from nearby mountains make Athens more comfortable than Sena Madureira.
(5) People in Athens are more accustomed to hot weather than are people in Sena Madureira.

Items 20 to 23 refer to the following article.

Something is killing bottleneck dolphins along the Atlantic coast. In less than 5 weeks, nearly 100 dolphins have washed ashore between New Jersey and Virginia. Experts claim that these dolphins represent only a handful of those who have died.

The cause of the dolphin deaths remains a mystery, although some clues have been uncovered. Pathologists who examined the bodies of the dead dolphins found evidence of bronchial pneumonia in all of them. Most of the dolphins also displayed body lesions that could have been caused by acid. This finding has drawn attention to the acid-waste dump located 100 miles off the coast of Cape May, New Jersey.

The director of the Marine Mammal Stranding Center in Brigantine, New Jersey, reports that the dolphin deaths have resulted in growing concern about ocean pollution.

"People call me," he said, "asking if what is killing the dolphins might pose a threat to them."

A scientist at an east coast research hospital stated, "We don't know what's killing the dolphins, but we do know that ocean water is polluted by sewage, chemical wastes, and heavy metals."

A doctor in Neptune, New Jersey, reports 300 cases of people developing infections after swimming in the ocean. The New Jersey Health Department has agreed to conduct water quality tests along the New Jersey shore.

20. According to the article, what is the main reason for concern about the dolphin deaths?

(1) Dolphins are an endangered species.
(2) Dolphins are essential to the economy of southern New Jersey.
(3) Harmful substances in ocean water might be killing the dolphins.
(4) Certain infections found in dolphins are contagious to human beings.
(5) People are reluctant to swim in the ocean because of the dolphin deaths.

21. Which of the following statements is a hypothesis that might explain why most of the dead dolphins have body lesions?

(1) Bronchial pneumonia has been found in the lungs of dead dolphins.
(2) Many people who swam in the same ocean water as the dolphins contracted infections.
(3) The ocean is being ruined by careless dumping of industrial wastes.
(4) The New Jersey Health Department is concerned about water quality along the New Jersey shore.
(5) Ocean water in the area contains acid pollutants.

22. Which of the following conclusions might be drawn from the fact that 300 swimmers have contracted infections after swimming in the same area in which dolphins have died?

(1) The same factors that are causing dolphins to die might be causing human beings to get sick.
(2) Both dolphins and human beings are susceptible to bronchial pneumonia.
(3) Human beings are more likely than dolphins to become ill from polluted ocean water.
(4) Human beings and dolphins are both sensitive to water contaminated by acid waste.
(5) Dead dolphins are polluting ocean water and causing human beings to get sick.

23. The California State Health Department discovers that large numbers of dead fish are washing ashore near San Diego. Based on the article, which of the following actions would be the most sensible first step for the department to take to find out why the fish are dying?

(1) Interview swimmers in the area to find out what they think of the water quality.
(2) Find out if an acid-waste dump exists near San Diego.
(3) Examine the water near San Diego to see if it contains chemical waste and heavy metal.
(4) Conduct autopsies on the bodies of the dead fish.
(5) Compare the number of dead fish to the number of infections contracted by swimmers in the area.

Item 24 refers to the following information and diagram.

The history of the Earth is divided into four major time periods called eras. The time line below shows the length and major events of each era.

Era	Length (millions of years)	
PRECAMBRIAN	4030	Seas form Mountains begin rising First life forms in the sea
PALEOZOIC	345	Volcanoes form First jawless fish First land plants First air-breathing animals First forests First amphibians and reptiles First insects
MESOZOIC	160	First dinosaurs First mammals First birds First flowering plants Palm & cone-bearing trees appear Dinosaurs die out
CENOZOIC	65	First horses Primates and human-like creatures develop Ice Age in North America First human beings Civilization begins

24. An earth science textbook contains five pictures that are described below. Which picture is a misrepresentation of the Earth's history?

(1) An early humanlike creature is shown riding a horse.
(2) The first air-breathing animals are shown roaming around the base of a volcano.
(3) The first organisms to dwell on Earth are shown rising up out of the ocean.
(4) A bird is shown sitting on the shoulder of a dinosaur.
(5) A dinosaur and a horse are shown in a stand of pine trees.

Item 25 refers to the following diagram.

Lunar Eclipse

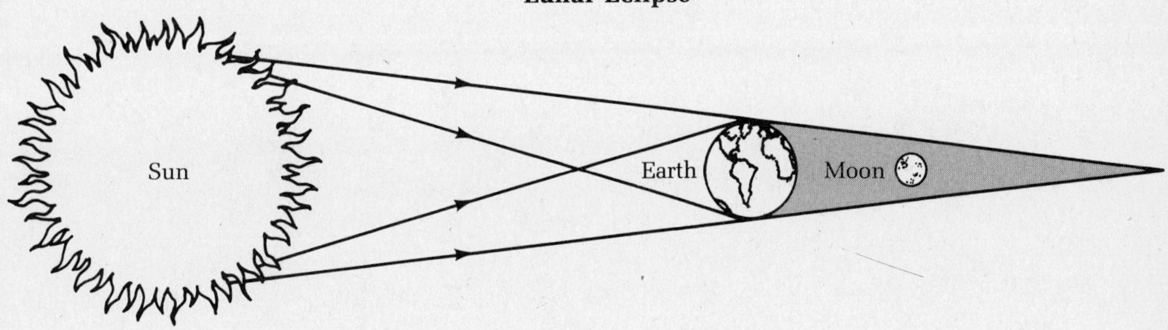

25. The diagram above shows a lunar eclipse, or eclipse of the moon. Which of the following statements explains why a lunar eclipse occurs?

(1) The moon is farther from the sun than usual.
(2) The northern hemisphere of the Earth is tilted away from the sun.
(3) The Earth casts a shadow on the moon, blocking out the sun's rays.
(4) The moon casts a shadow on the Earth, blocking the sun's light.
(5) The Earth's rotation causes half of the Earth to be in darkness while the other half is in light.

Item 26 refers to the following graphs.

The graphs below show the changes in pressure and temperature that occur below the Earth's surface.

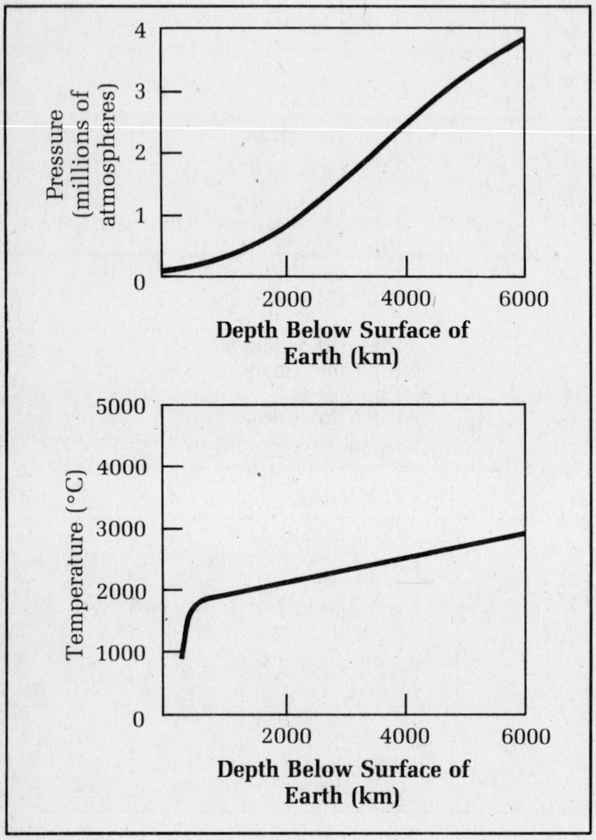

26. According to the graphs, as distance from the Earth's surface increases,

(1) temperature and pressure increase at approximately the same rate.
(2) temperature increases sharply, whereas pressure remains about the same.
(3) temperature increases sharply, whereas pressure decreases.
(4) temperature increases most sharply near the center of the Earth, whereas pressure increases most sharply just below the Earth's surface.
(5) temperature increases most sharply just below the Earth's surface, whereas pressure increases least sharply near the Earth's surface.

Item 27 refers to the following information and diagram.

The experimental setup shown below was designed by an earth science student who wanted to investigate the heating of the Earth by the sun.

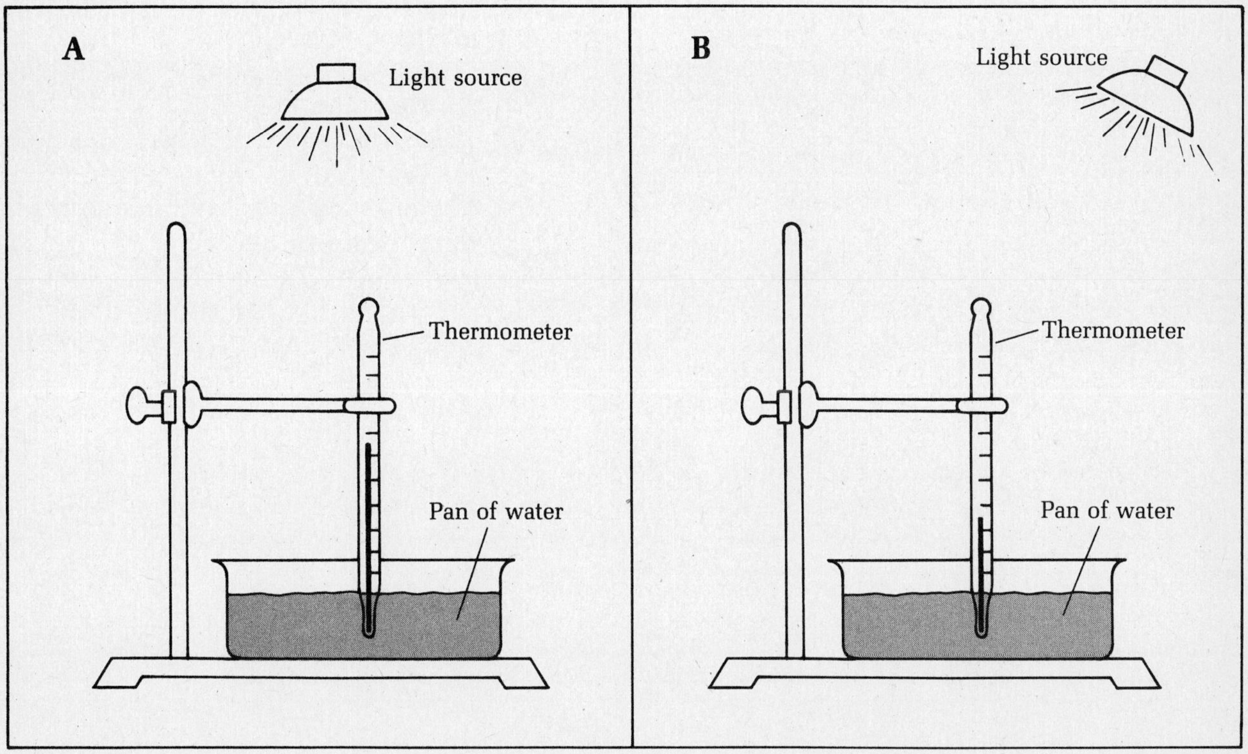

27. If the light source represents the sun and the pan of water represents the Earth, which of the following hypotheses would be tested in this experiment?

(1) Areas with large bodies of water are generally cooler than other areas.
(2) Areas with large bodies of water are generally warmer than other areas.
(3) The distance of the Earth from the sun accounts for the variation in temperature between summer and winter.
(4) Areas heated by direct rays of the sun are warmer than areas heated by indirect rays of the sun.
(5) Temperatures generally are warmer in the southern hemisphere than in the northern hemisphere because the southern hemisphere receives more sunlight.

Item 28 refers to the following information and chart.

Climate types often are classified according to the P/Ep ratio. This ratio compares the average yearly precipitation with the average yearly evapotranspiration. Evapotranspiration refers to the total water loss by direct evaporation as well as by transpiration, or the giving off of moisture through the surfaces of leaves and other plant parts. Evapotranspiration is directly related to temperature—the higher the temperature, the higher the Ep value.

Climate Type	P/Ep
Arid	less than 0.4
Semiarid	0.4–0.8
Subhumid	0.8–1.2
Humid	greater than 1.2

28. Regions A and B receive the same amount of yearly rainfall, yet Region A is semiarid, whereas region B is subhumid. Which of the following statements explains why?

(1) Region A experiences higher temperatures than Region B.
(2) Region A experiences lower temperatures than Region B.
(3) Region A is located closer to the equator than Region B.
(4) Region A has more desert land than Region B.
(5) Region A has more mountain ranges than region B.

Items 29 to 32 refer to the following information and diagram.

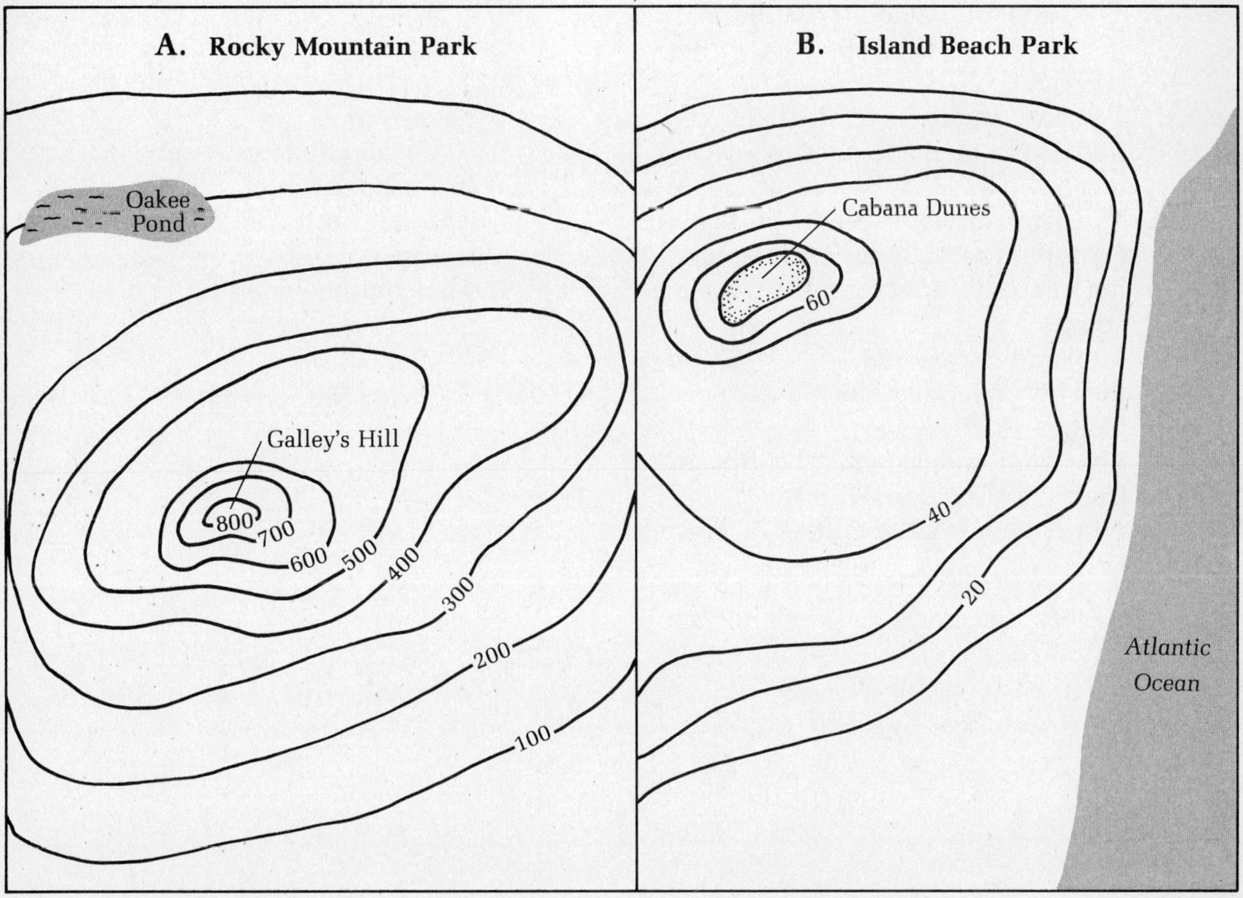

A topographic map shows the relief, or differences in elevation, of an area. On a topographic map, changes in elevation are shown by the use of contour lines. The distance between contour lines is called the contour interval. In the topographic map shown in Figure A, the contour interval is 100 feet.

29. What is the difference in elevation between Galley's Hill and Oakee Pond in Figure A?

(1) 800 ft.
(2) 600 ft.
(3) 300 ft.
(4) 200 ft.
(5) 100 ft.

30. What is the contour interval on the map in Figure B?

(1) 10 ft.
(2) 20 ft.
(3) 40 ft.
(4) 60 ft.
(5) 80 ft.

31. How would you describe the land surface in Figure B?

(1) The relief of the land is varied, with flat areas and steep areas interspersed.
(2) The land surface is totally flat, rising little more than a few feet above sea level.
(3) The land surface is like a great plateau, flat for many miles, then sloping sharply down to the water's edge.
(4) The land slopes gently upward from the shoreline, with the highest point being a small hill.
(5) The land slopes sharply upward from the shoreline, with the highest point being a sharp cliff.

32. What is the most likely explanation when contour lines on a topographic map are very close together?

(1) The map is drawn to small scale.
(2) The elevation of the land does not change much.
(3) The land is more densely populated in this area.
(4) The elevation of the land rises sharply.
(5) The contour interval is small.

Check your answers on pages 124–126.

Chemistry

Directions: Choose the one best answer to each of the following questions.

Items 1 and 2 refer to the following chart.

BOILING POINTS OF WATER AT VARIOUS ATMOSPHERIC PRESSURES	
Atm. Pressure (mm mercury)	**Boiling Point (° Celsius)**
760 (standard)	100
724	98
778	101

1. What is the relationship between atmospheric pressure and the boiling point of water?

(1) The boiling point of water is independent of atmospheric pressure.
(2) The boiling point of water increases as atmospheric pressure decreases.
(3) The boiling point of water increases as atmospheric pressure increases.
(4) Water boils most rapidly at standard pressure.
(5) Water boils at the same temperature regardless of atmospheric pressure.

2. Atmospheric pressure decreases with altitude. With this in mind, why does it take longer to cook an egg on a mountaintop than at sea level?

(1) Water takes longer to boil on a mountaintop than at sea level.
(2) An egg cooks more slowly at a lower temperature than at a higher temperature.
(3) Water boils more rapidly at sea level than on a mountaintop.
(4) An egg cooks most rapidly at standard atmospheric pressure.
(5) Water molecules move more slowly at high altitudes than at low altitudes.

Items 3 and 4 are based on the following passage.

The properties of a substance are definite characteristics by which the substance can be recognized. For example, some important properties of ice are that it is solid, cold, and melts at 0°C.

The properties of a substance can be classified as either physical or chemical. Physical properties are those characteristics that can be observed or measured without changing the nature of the substance. Color, odor, taste, density, hardness, boiling point, and melting point are common physical properties.

Chemical properties describe the way a substance reacts or fails to react in the presence of other substances. For example, the ability of iron to combine with oxygen and water vapor to form rust is a chemical property of iron. The inability of neon gas to react with most other elements is a chemical property of neon.

3. Which of the following situations describes a chemical property?

(1) TNT explodes to produce CO_2, water vapor, and nitrogen.
(2) Iron becomes molten when exposed to very high temperatures.
(3) Mercury becomes a liquid at room temperature.
(4) Bubbles of carbon dioxide gas rise to the surface of water because they are less dense than water.
(5) Ethylene glycol prevents water in a car radiator from freezing in cold weather.

4. In which of the following situations would knowledge of the properties of a substance be most useful?

(1) determining how much cloth is needed to upholster a chair
(2) determining whether a gold-colored ring is really made of gold
(3) deciding how much flour should be added to bread dough
(4) determining which model car gets the best mileage
(5) determining how long to cook a steak in a microwave oven

Item 5 refers to the following information.

Indirect evidence is information obtained about something that cannot be observed or measured directly. Scientists have had to use indirect evidence to learn about the structure of the atom, because an atom is too small to be seen. Much of what scientists know about the atom has come from observing the light energy given off by certain substances and by observing the electrical behavior of certain substances.

5. Which of the following situations involves the use of indirect evidence?

A. A child shakes a wrapped birthday present, hoping to determine what is inside.
B. A man places his spare tire in a tub of water to see if air bubbles indicate a leak.
C. A woman measures a piece of cloth to see if it is large enough to cover her sofa.

(1) A only
(2) B only
(3) C only
(4) A and B only
(5) B and C only

Item 6 is based on the following information.

When milk reacts with an acid, a coagulated protein is formed. Sour milk is a common example of milk with coagulated protein. When milk is mixed with vinegar, a curd forms and the milk tastes sour.

Three chemistry students draw the following conclusions from the information given:

A. Vinegar is a coagulated protein.
B. Vinegar is an acidic substance.
C. The curd in milk is a coagulated protein.

6. Which of the above conclusions is (are) supported by the information?

(1) A only
(2) B only
(3) C only
(4) A and B
(5) B and C

Item 7 is based on the following information.

A molecule of sucrose, or common table sugar, contains carbon (C), oxygen (O), and hydrogen (H). The formula for sucrose is

$$C_{12}H_{22}O_{11}$$

7. According to the above information, a molecule of sucrose contains

(1) equal amounts of carbon, oxygen, and hydrogen.
(2) twice as many carbon atoms as hydrogen and oxygen atoms.
(3) twice as many hydrogen atoms as oxygen atoms.
(4) more oxygen atoms than carbon or hydrogen atoms.
(5) equal amounts of carbon and oxygen.

---

8. Silver tarnishes when it comes into contact with sulfur. Because egg yolk contains hydrogen sulfide (H_2S), silver tarnishes quickly when exposed to egg yolk. A silver spoon also tarnishes when a rubber band is wrapped around it. Which of the following would explain best why silver tarnishes when wrapped with a rubber band?

(1) Rubber and silver react to form hydrogen sulfide.
(2) Rubber bands are made of the same substance as egg yolks.
(3) The tight rubber band puts pressure on the silver.
(4) Rubber bands contain some form of sulfur.
(5) Rubber speeds up the reaction between silver and sulfur.

Items 9 to 12 refer to the following passage.

According to the Bohr model of the atom, electrons can be found in rings around the nucleus. These rings are called energy levels or shells and are referred to by number.

Level 1 is closest to the nucleus and contains electrons with the least amount of energy. This level can hold a maximum of 2 electrons. Level 2 is next closest to the nucleus and can hold a maximum of 8 electrons. The third energy level can hold up to 18 electrons, and the fourth energy level can hold as many as 32 electrons.

In any atom, the energy level farthest from the nucleus is called the outer energy level, or outer shell. It is the electrons in the outer shell that participate in chemical reactions. For this reason, the chemical properties of an element are closely related to the number of electrons in the outer shell.

9. An atom of sodium contains 11 electrons. Assuming that electrons fill energy levels in order, beginning with level 1, which energy level would be the outer energy level in the sodium atom?

(1) level 1
(2) level 2
(3) level 3
(4) level 4
(5) level 2 or level 3

10. An atom of nitrogen contains 7 electrons. How many electrons would be present in the outer energy level of a nitrogen atom?

(1) 0
(2) 2
(3) 5
(4) 7
(5) 8

11. Beryllium, magnesium, and calcium exhibit similar behavior in chemical reactions. Based on the passage, which of the following statements is the most likely explanation for this phenomenon?

(1) All three substances are elements.
(2) All three substances are metals.
(3) All three substances are in the same column in the Periodic Table.
(4) All three substances have 2 electrons in their outer shells.
(5) All three substances have the same number of electrons as protons.

12. Under certain conditions, an electron can gain enough energy to jump from one energy level to another. Which of the following conditions would have to be met in order for an electron to jump from the second energy level to the third energy level?

(1) The second energy level would have to contain fewer than 8 electrons.
(2) The third energy level would have to contain fewer than 18 electrons.
(3) The first energy level would have to be full.
(4) The third energy level would have to be full.
(5) The third energy level would have to contain fewer electrons than the second energy level.

Item 13 is based on the following information and diagram.

The diagram shows how to read the volume of a liquid in a graduated cylinder. The curved surface of a liquid in a vessel is called the meniscus.

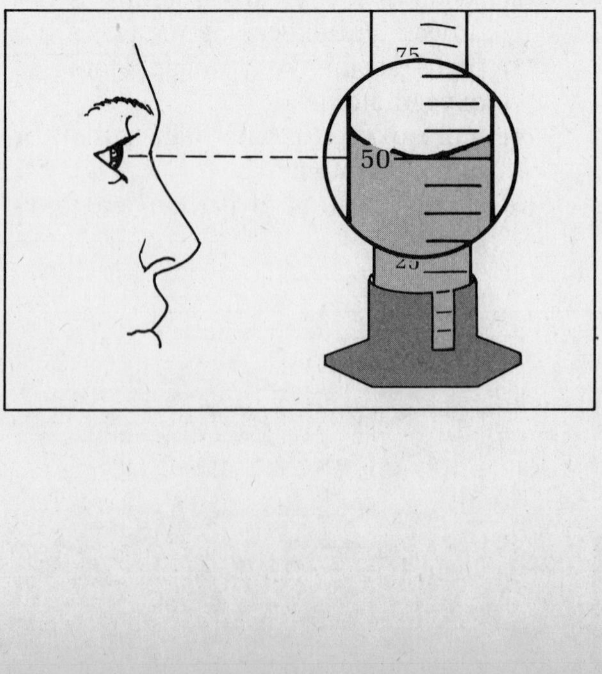

13. If you were following instructions to mix a particular shade of wood stain, what would be the correct way to read the amounts you were mixing?

(1) Look straight down into the graduated cylinder and note the level of the liquid.
(2) View the surface of the liquid at eye level and note the level of the liquid at the outer edge of the meniscus.
(3) View the graduated cylinder from slightly below the meniscus and note the level of the liquid at the lowest portion of the meniscus.
(4) View the surface of the liquid at eye level and note the level of the liquid at the lowest portion of the meniscus.
(5) View the surface of the liquid at eye level and take the average of the levels of the liquid at the highest and lowest portions of the meniscus.

Items 14 to 17 refer to the following information.

"Relief is just a swallow away with Tum-Aid! Tum-Aid relieves acid indigestion faster than any other leading stomach remedy. Tum-Aid's secret ingredient gobbles up excess stomach acid and has you feeling better in a minute. On the job or on the go, you can trust your tummy to Tum-Aid!"

14. An acid and a base react to produce a salt and water. The "secret ingredient" in Tum-Aid is most likely

(1) water.
(2) an acid.
(3) a salt.
(4) a base.
(5) an enzyme.

15. Which of the following sales pitches is backed up by scientific evidence?

(1) Tum-Aid tastes better than other leading brands.
(2) Tum-Aid neutralizes stomach acid.
(3) Tum-Aid contains a secret ingredient.
(4) Tum-Aid makes you feel better faster.
(5) Tum-Aid is the late-night snacker's best friend.

16. Which of the following is most likely to be a side effect of taking Tum-Aid?

(1) dehydration
(2) heartburn
(3) drowsiness
(4) excessive acid in the small intestines
(5) increased sodium in the bloodstream

17. A person takes Tum-Aid yet fails to obtain relief from his indigestion. According to the information provided, which of the following statements could explain why this happened?

(1) The person failed to follow the directions on the package.
(2) The person took too little Tum-Aid to obtain relief.
(3) The person's stomach problem is not the result of excess stomach acid.
(4) The person has a natural immunity to this type of medication.
(5) The person purchased a package of Tum-Aid that had been tampered with.

Item 18 refers to the following information and diagram.

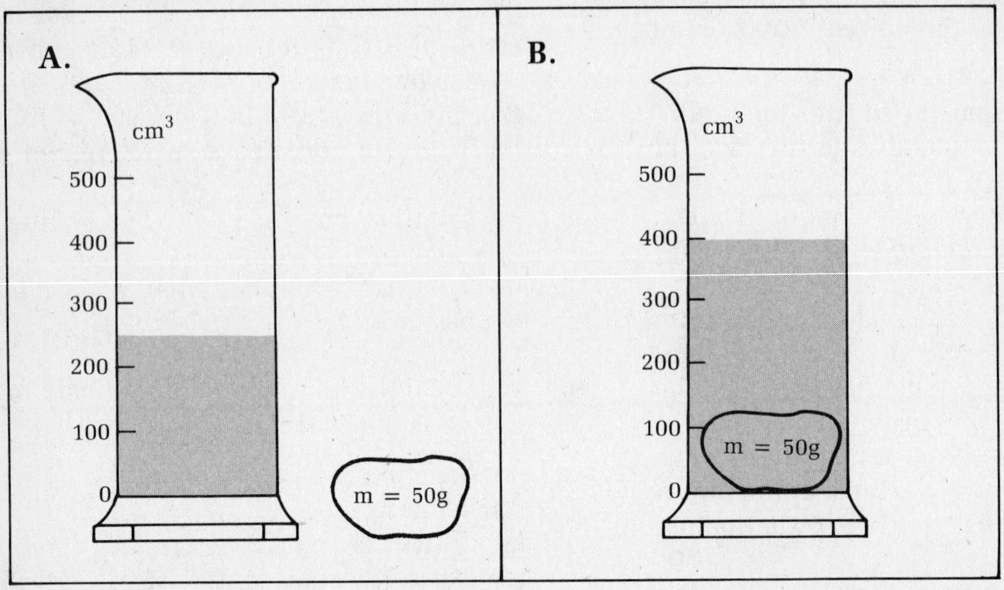

When immersed in water, a substance will displace a volume of water equal to its own volume.

18. What is the volume of the rock shown in the diagram above?

(1) 50 g
(2) 150cm^3
(3) 250cm^3
(4) 400cm^3
(5) 500cm^3

Item 19 refers to the following information and diagram.

When a flashbulb is set off, the white powder magnesium oxide is produced. The diagram below shows how a flashbulb looks before this reaction occurs.

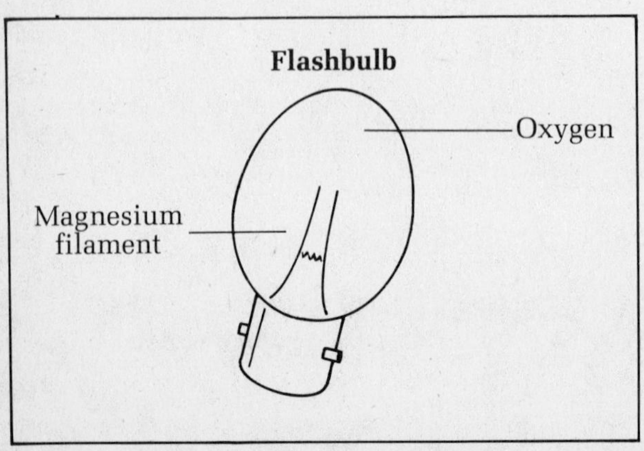

19. Based on the information provided, which of the following statements could explain why a flashbulb appears dented and slightly smaller after it has been used?

(1) The heat from the light flash melts the bulb.
(2) Consumption of the oxygen in the bulb creates a vacuum, causing the bulb to collapse somewhat.
(3) Particles of magnesium oxide push against the bulb, causing the bulb to dent.
(4) The explosive force of the chemical reaction inside the bulb causes the bulb to collapse.
(5) The combination of heat and light produced by the flash causes the bulb to decompose.

Items 20 to 23 refer to the following passage.

The rate of a chemical reaction is affected by four factors. These factors are concentration, surface area, temperature, and the presence of a catalyst.

The concentration of a substance is a measure of the amount of that substance in a given unit of volume. A high concentration means that there are many particles per unit of volume. As a result, substances present at high concentrations tend to react more rapidly than substances present at low concentrations.

When one or more of the participating substances in a chemical reaction is a solid, the rate of reaction can be increased by breaking the solid into smaller pieces. This increases the surface area of the solid, which makes it easier for the particles of the substances to come into contact and react.

An increase in temperature causes the particles of a substance to gain energy and move more rapidly. As a result, these particles are more likely to collide with each other and react. Most chemical reactions show an increase in rate as temperature increases. Many chemical reactions double or triple in rate with a rise in temperature of 10°C.

Some chemical reactions take place very slowly—much too slowly to be of practical value. Such reactions can be speeded up by the addition of a catalyst. A catalyst is a substance that increases the rate of reaction but is not itself changed by the reaction.

20. A log will take many hours to burn completely, but if the same log is made into sawdust, the sawdust will burst into flames explosively when ignited with a match. Which of the following statements explains why this happens?

(1) Wood is present at a higher concentration in sawdust than in a log.
(2) Wood is present at a lower concentration in sawdust than in a log.
(3) The match acts as a catalyst in igniting the sawdust.
(4) The wood in sawdust has a greater surface area than the wood in a log.
(5) The temperature is greater when a match is thrown into sawdust than when a log burns slowly.

21. Which of the following suggestions would be the best way to cook a piece of meat rapidly?

(1) Cut the meat into several pieces.
(2) Coat the meat with barbecue sauce.
(3) Place the meat in a covered dish.
(4) Cook the meat over a low flame.
(5) Combine the meat with other ingredients.

22. A lab technician is instructed to combine chemicals A and B in a small beaker. When she mixes the chemicals together, she finds that the reaction is so violent that the product splatters out of the beaker. In order to stop the splattering, the technician should

(1) place the beaker on a hot plate.
(2) add a catalyst to the mixture.
(3) place the beaker in a bucket of ice.
(4) increase the concentration of Chemical A.
(5) increase the concentration of Chemical B.

23. In the human body, enzymes function as catalysts. Which of the following situations might occur if a person failed to produce enzymes?

(1) Loud noises would become more irritating.
(2) The person's heartbeat would increase.
(3) The person would suffer from periods of insomnia.
(4) A tap on the knee would produce a more rapid reflex than normal.
(5) A piece of food would take weeks to digest.

Item 24 is based on the following information.

Density is the mass per unit of a substance. When a gas is heated, the molecules in the gas gain energy and begin to move more rapidly. They spread farther apart and take up a larger volume than before. As a result, the density of a gas decreases as temperature increases. If two gases of unequal densities are mixed together, the gas that is less dense will rise above the gas that is more dense.

24. A person plans to build a house with a central heating and air-conditioning system. Which of the following describes the most practical plan for placing the heating and air-conditioning vents in each room?

(1) Place the heating vents near the ceiling and the air-conditioning vents near the floor.
(2) Place the heating vents near the floor and the air-conditioning vents near the ceiling.
(3) Place both the heating and air-conditioning vents about halfway between the floor and the ceiling.
(4) Place the heating and air-conditioning vents as far apart as possible.
(5) Place the heating and air-conditioning vents directly across from each other.

Check your answers on pages 126–128.

Physics

Directions: Choose the one best answer to each of the following questions.

Items 1 to 3 refer to the following graph.

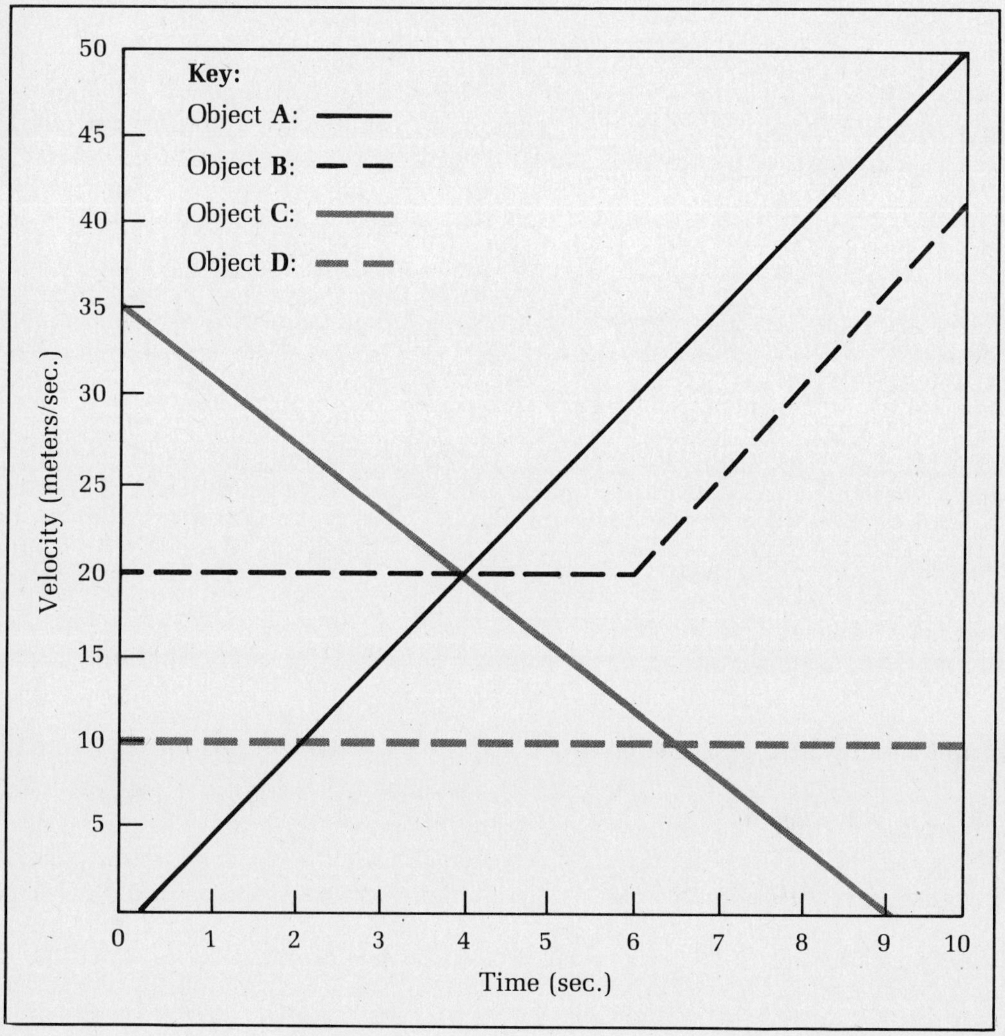

Velocity is the speed of an object in a given direction. Acceleration is the rate of change in velocity.

1. According to the graph, which of the following statements is true about object D?

   (1) It is standing still.
   (2) It is coming to a stop.
   (3) It is accelerating.
   (4) It is moving at a constant velocity.
   (5) It is moving to the right.

2. According to Newton's first law of motion, an object at rest tends to remain at rest, and an object in motion tends to continue in motion in a straight line, unless acted on by an outside force. Which of the objects must have been acted on by an outside force during the first 4 seconds shown on the graph?

   (1) A only
   (2) B only
   (3) A and C
   (4) C and D
   (5) A, B, C, and D

3. During the last 4 seconds, which of the following statements is true about objects A and B?

   (1) They are moving side by side.
   (2) They both are moving at a constant speed.
   (3) They both are moving at the same speed.
   (4) They are moving in opposite directions.
   (5) They both are accelerating at the same rate.

4. The mass per unit of volume of an object is called its density. Density can be expressed by the following mathematical formula:

   density = mass/volume

   Which of the following statements correctly describes the relationship between a kilogram of feathers and a kilogram of lead?

   (1) The feathers and the lead have the same mass, but the feathers take up more space.
   (2) The feathers and the lead have the same mass, but the lead is heavier.
   (3) The feathers take up more space, but the lead has more mass.
   (4) The feathers have more mass, but the lead has a greater density.
   (5) The feathers and the lead take up the same amount of space, but the lead is heavier.

Items 5 to 7 refer to the following passage.

Movement can be measured only with reference to something that is assumed to be fixed in place. The object or point from which movement is determined is called a frame of reference.

The most common frame of reference is the Earth. When a plane flies overhead or a car passes by, its speed is measured relative to the surface of the Earth. Yet the Earth is not really still. The Earth rotates from west to east on its axis at a speed of 16.7 kilometers per hour. Compared to a fixed point in space, all objects "at rest" on the Earth's surface are moving at this rate.

Perhaps you have had the experience of riding in a train and being passed by another train going slightly faster. Suppose your train is moving at 90 km/hr and the faster train is moving at 100 km/hr. To an observer standing near the railroad tracks, both trains appear to be moving very rapidly, at speeds of 90 and 100 km/hr. To you, however, the faster train appears to be moving only 10 km/hr.

5. Which of the following is the best reason for choosing the Earth as a frame of reference?

(1) It is necessary to have a standard reference point for the measurement of everyday movement.
(2) The Earth's rotation compensates for any inaccuracies in the measurement of movement.
(3) The Earth is stationary compared with other objects in space.
(4) Movement usually appears the same, no matter what frame of reference it is viewed from.
(5) A speed of 16.7 km/hr is a convenient standard for measuring movement.

6. A car travels from east to west across the United States at a speed of 90 km/hr. Relative to a fixed point in space, how fast is the car traveling?

(1) 73.3 km/hr
(2) 90 km/hr
(3) 106.7 km/hr
(4) 167 km/hr
(5) 1,503 km/hr

7. A person sitting in a stationary bus feels that the bus is moving backward as the car next to it begins to move forward. Based on the passage, what best explains this?

(1) A defect in the bus parking brake causes the bus to roll backward.
(2) As the car pulls forward, it accidentally bumps the bus, causing the bus to roll backward.
(3) Because of the Earth's rotation, the bus appears to be moving backward.
(4) If the moving car is used as a frame of reference, the bus appears to be moving backward away from the car.
(5) People walking on the sidewalk in the same direction as the car become the new frame of reference and make the bus appear to be moving.

Items 8 to 10 refer to the following passage.

A wave is a rhythmic disturbance that transfers energy from one place to another. Some common types of waves include ocean waves, sound waves, light waves, and radio waves.

The substance through which a wave travels is called a medium. Water is the medium for ocean waves; air is a common medium for sound, light, and radio waves. Not all waves require a medium. Light waves and radio waves can travel through a vacuum as well as through solids, liquids, and gases. Sound waves, on the other hand, must have a medium through which to travel.

Different types of waves travel at different speeds. For example, sound waves travel in air at a speed of 346 meters per second, whereas light waves travel at a speed of 300,000 kilometers per second. The medium through which a wave travels also affects the speed of the wave. Sound waves show one of the greatest variations in speed, traveling up to 6,000 meters per second through solids and up to 1,500 meters per second through liquids. Sound waves travel most slowly through gases.

8. A person in an apartment building receives a complaint from his next-door neighbor about the volume of his television set. Based on the passage, which of the following actions might reduce the noise in his neighbor's apartment?

(1) Place a rug under the television set.
(2) Pull the television set several feet out from the wall.
(3) Make sure that the television set is firmly touching the wall.
(4) Change the type of speaker used in the television set.
(5) Place a piece of cloth over the television set.

9. Thunder is the sound that is produced by lightning. If lightning strikes very close by, a person will hear the thunder at the same time the lightning is seen. The farther the person is from the lightning strike, the longer it will be before the thunder is heard. Which of the following statements explains why this happens?

(1) Light waves can travel through a vacuum, whereas sound waves cannot.
(2) Sound waves can travel through a medium, whereas light waves cannot.
(3) Sound waves travel more slowly through gases than through solids.
(4) Light waves can pass through gas molecules, but sound waves cannot.
(5) Light waves travel more rapidly than sound waves.

10. Based on this passage, which of the following explains why music from a radio could not be heard on the moon?

(1) Radio waves cannot travel far enough to reach the moon.
(2) The moon has no atmosphere.
(3) Gravity on the moon is only one sixth as strong as gravity on the Earth.
(4) Sound waves cannot travel fast enough to reach the moon.
(5) There is no water on the surface of the moon.

Items 11 to 14 refer to the following passage.

Many people think of work as mowing the lawn, washing dishes, or doing a job for pay. But to a scientist, work is done when a force acts on an object and causes it to move.

Energy is the capacity for doing work. All energy can be classified into two types: potential and kinetic. Potential energy is the energy a body has because of its position or state. It is stored energy, waiting for a force to release it.

An object resting above ground has potential energy because of gravity. As long as the object rests in the same place, it does no work. But if the object is allowed to fall, it can do work. For example, a cement block might fall from a roof and hit the top of a stake, driving it deeper into the ground.

Kinetic energy is the energy of motion. All moving objects have kinetic energy. Different forms of energy are included in kinetic energy. Because light energy, sound energy, and heat energy all move from one place to another, they are all kinetic energy. The amount of kinetic energy an object has is related to its mass and its velocity.

Potential energy can be converted into kinetic energy without any loss of energy. A log sitting by the fireplace has potential energy as long as it remains unused. When it is used to make a fire, the potential energy is converted to heat, a kind of kinetic energy.

Until the twentieth century, scientists believed that energy could be neither created nor destroyed. This belief was called the law of conservation of energy. Today scientists know that energy can be created from matter during nuclear reactions. As a result, the law of conservation of energy has been restated as the law of conservation of matter and energy.

11. Which of the following situations does not involve work?

(1) turning on a light switch
(2) lifting a child out of a crib
(3) standing in line while holding a bag of groceries
(4) handing a quarter to a toll collector
(5) typing a term paper

12. Which of the following scientific facts is evidence that scientists change laws on the basis of new information?

(1) If the mass and velocity of an object change, so does its amount of kinetic energy.
(2) Matter can change into energy during a nuclear reaction.
(3) Energy can be neither destroyed nor created.
(4) Potential energy can be converted into kinetic energy.
(5) Heat, light, and sound are all kinds of kinetic energy.

13. Water at the bottom of a waterfall is usually several degrees warmer than water at the top. Based on the passage, which of the following statements explains why this is so?

(1) A type of kinetic energy warms the water as it moves over the falls.
(2) Because water moves faster at the bottom of the falls than at the top of the falls, the temperature of the water rises.
(3) As the water falls, its mass increases, causing some of its potential energy to change into heat energy.
(4) Kinetic energy is greater at higher elevations than at lower elevations, whereas heat energy is greater at lower elevations than at higher elevations.
(5) Work done on the water causes its temperature to rise.

14. A woman trades her 8-lb. bowling ball for a 12-lb. bowling ball. Which of the following best explains why, when she rolls the ball as usual, the 12-lb. ball travels more slowly than the 8-lb. ball?

(1) It takes more work to move a 12-lb. ball than to move an 8-lb. ball.
(2) It takes more skill to roll a 12-lb. ball than to roll an 8-lb. ball.
(3) There is more friction between the alley and a 12-lb. ball than between the alley and an 8-lb. ball.
(4) The 8-lb. ball has a smoother surface than the 12-lb. ball.
(5) An 8-lb. ball has more momentum than a 12-lb. ball.

Items 15 and 16 are based on the following information and diagram.

Because light travels at different speeds in different mediums, a beam of light is refracted, or bent, as it passes from one medium into another. The size of the angle of refraction depends on the particular medium into which the light is going.

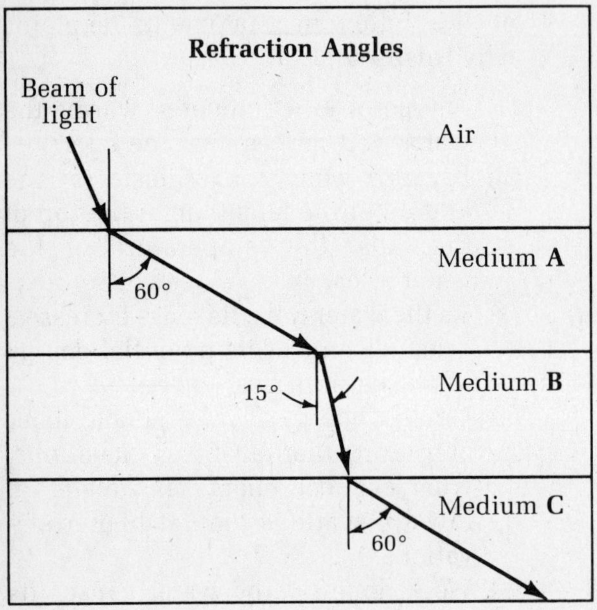

15. According to the diagram, which of the following best supports the conclusion that mediums A and C are the same?

(1) Mediums A and C are both different from medium B.
(2) Light is reflected from medium A into medium C.
(3) Mediums A and C have the same density.
(4) Light does not change speed as it passes from medium A into medium C.
(5) Mediums A and C show the same angles of refraction.

16. Suppose that medium A is water. According to the way light behaves as it passes from air to water, which of the following statements will be true about a silver dollar sitting at the bottom of a shallow pond?

(1) The silver dollar would look larger than it really is.
(2) The silver dollar would look smaller than it really is.
(3) The silver dollar would appear to be several inches away from the place where it really is.
(4) The silver dollar would appear closer to the surface of the pond than it really is.
(5) The silver dollar would appear to be a different color than it really is.

Item 17 is based on the following information and diagram.

The force required to accelerate an object is equal to the product of the object's mass times the acceleration of the object (F = ma).

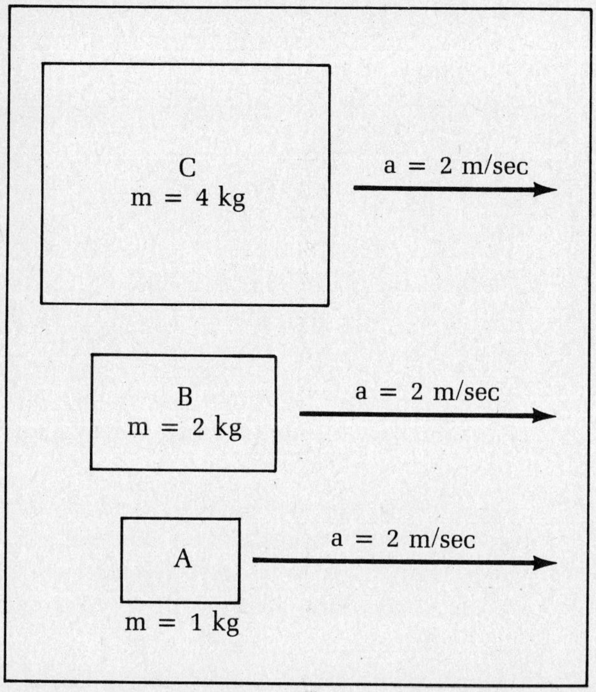

17. Which of the following statements must be true about the forces that produced the motion of objects A, B, and C?

(1) All the forces must have been equal.
(2) The force acting on object A must have been twice as great as the force acting on object B.
(3) The force acting on object C must have been twice as great as the forces acting on objects A and B.
(4) The force acting on object A must have been twice as great as the force acting on object B and four times as great as the force acting on object C.
(5) The force acting on object C must have been twice as great as the force acting on object B and four times as great as the force acting on object A.

Items 18 to 20 refer to the following passage.

Radioactivity is the release of matter and energy from an atomic nucleus. Many elements are naturally radioactive. Among these are uranium, radium, and a certain form of carbon. Other radioactive elements can be created in the laboratory. Some of these include forms of iron, iodine, and sodium.

Radioactivity cannot be seen or felt. For this reason, special instruments must be used to detect it. One of the best-known detectors of radioactivity is the Geiger counter. In the presence of a radioactive substance, an electric current is produced in the Geiger counter and light flashes or clicks are emitted. The number of flashes or clicks per unit of time indicates the strength of the radioactive substance.

The detection of radioactivity is important, because radioactivity in large amounts is harmful to living things. When an organism is exposed to too much radiation, its cells can be damaged or destroyed and the entire organism might die.

18. Based on the passage, which of the following would not be a consequence of exposure to radiation?

(1) A woman exposed to radiation gives birth to a child with abnormal cell structures.
(2) Radioactive materials are dumped into a lake, and many of the fish and plants die.
(3) A laboratory animal exposed to radiation suffers brain damage and loses some of its normal functions.
(4) Plants grown in soil polluted by radioactive materials show an increased number of chloroplasts and other food-producing cells.
(5) A scientist who handles radioactive materials develops a degenerative muscle disease.

19. For which of the following purposes would a Geiger counter be useful?

(1) detecting an electric current
(2) detecting the location of a radioactive substance
(3) determining the identity of a radioactive substance
(4) determining the half-life of a radioactive substance
(5) determining the origin of a radioactive substance

20. People who work with radioactive materials often wear radio-sensitive badges. Based on the passage, which of the following statements could explain why these badges are worn?

(1) The badges help to identify workers who have special knowledge about radiation.
(2) The badges are useful to scientists who want to measure radiation.
(3) The badges help identify the radioactive substances that the people are handling.
(4) The badges are useful in testing the effectiveness of Geiger counters.
(5) The badges serve as a warning if the workers are exposed to too much radiation.

Items 21 and 22 are based on the following information and chart.

The specific heat of a substance is the amount of heat required to raise one gram of the substance one degree Celsius. The table below shows the specific heats of some common stubstances.

Substance	Specific Heat (Cal/g-C°)
Air	0.25
Alcohol	0.594
Aluminum	0.22
Glass	0.20
Gold	0.030
Iron	0.107
Lead	0.030
Water	1.0
Wood	0.42

21. A factory machine that produces steel bearings is cooled by a container of water that absorbs heat from the metal parts of the machine. A technician notices that the machine is consistently overheating. She considers three possible actions:

A. Increase the mass of the steel bearings.
B. Increase the size of the container of water.
C. Replace the iron parts of the machine with parts made of aluminum.

Which of the above actions would effectively stop the machine from overheating?

(1) A only
(2) B only
(3) A or B
(4) B or C
(5) A, B, or C

22. A person who lives near a large body of water notices that on a hot day in early summer, the water temperature is considerably lower than the temperature of the surrounding air. Based on the information provided, which of the following statements explains why this is so?

(1) Cool breezes blowing across the water make the water seem cooler.
(2) Heat from the water has escaped into the atmosphere.
(3) Trees overhanging the water provide shade that keeps the water cool.
(4) Water can absorb more heat than air without undergoing a change in its own temperature.
(5) Frequent rainfall in the area has replaced warmer water with cooler water.

Item 23 is based on the following information and diagram.

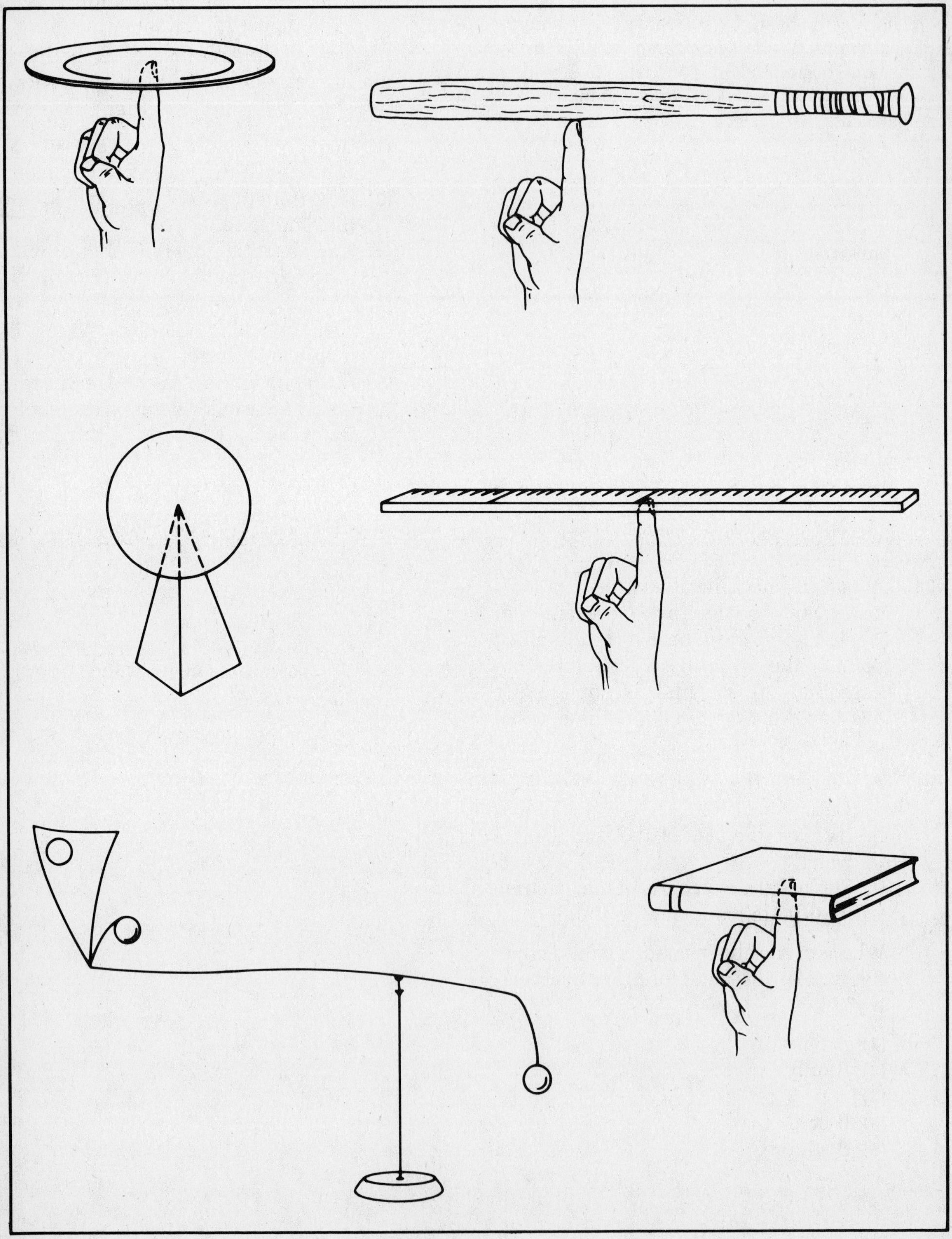

The center of gravity of an object is the point at which all of the object's mass appears to be concentrated. An object is in balance when its center of gravity is supported. The diagram below shows several different objects being balanced at their centers of gravity.

23. Which of the following statements best describes the relationship between the shape of an object and its center of gravity?

(1) The center of gravity in an object of uniform shape will be at the center of the object, whereas the center of gravity in an object of irregular shape will not be at the center of the object.
(2) The center of gravity in an object will be at the center of the object, regardless of its shape.
(3) The center of gravity in a rectangular object always will be closer to one end of the object, whereas the center of gravity in a round object always will be at the center of the object.
(4) The center of gravity in an object varies according to the size of the object.
(5) The center of gravity in an object of uniform shape might be at the center of the object, whereas the center of gravity in an object of irregular shape is always at the center of the object.

Item 24 is based on the following information.

Most substances expand when heated. One exception to this rule is water. At temperatures near the freezing point, water expands as it is cooled.

24. An early frost catches a backyard gardener off guard. When she comes out in the morning to inspect her garden, she finds several of her favorite ceramic planters cracked and lying in pieces around the hard root balls of the plants. Which statement below explains what happened?

(1) The roots grew quickly in response to the cold in an attempt to obtain extra nutrients to last through the winter.
(2) The pots contracted because of the cold and cracked.
(3) The water in the soil expanded and cracked the pots.
(4) Frost on the pots caused them to crack.
(5) The change in temperature from cold to warm when the sun came up caused the pots to crack.

Check your answers on pages 128–129.

HALF-LENGTH PRACTICE TEST

Directions: Choose the one best answer to each of the following questions.

1. Various things happen to sunlight when it reaches the Earth's surface. Some of the light is absorbed by dark substances, such as tar. Some passes through clear substances, such as glass or shallow water. The rest is reflected by shiny objects, such as metal.

According to this information, which color would be best for solar energy-collecting panels, such as those found on the roof of a solar-powered home?

(1) white
(2) metallic silver
(3) black
(4) blue
(5) red

2. Humidity is the amount of water vapor held by a given volume of air. Relative humidity is the ratio of the amount of water the air is holding compared to the maximum amount of water it could hold at a given temperature. Warmer air can hold more water than the same amount of cooler air. At which of the following times of day is the relative humidity normally the lowest?

(1) midnight
(2) 2 a.m.
(3) 8 a.m.
(4) noon
(5) 10 p.m.

Items 3 to 7 are based on the following information.

Organisms can be classified according to how they relate to each other. Below are five types of relationships classified according to the benefit or harm to each organism.

A. mutualism:	different organisms live together with all benefiting from the relationship
B. commensalism:	benefits one partner only; the other is neither harmed nor helped by the relationship
C. parasitism:	benefits the parasite only; the host is harmed by the relationship but usually lives
D. predation:	benefits the predator only; the prey is killed
E. saprophytism:	the organism feeds on dead organisms or the nonliving products of living organisms to its benefit

Each of the above categories might be used more than once in the following items.

3. Nitrogen-fixing bacteria live in root nodules of pea plants. The bacteria obtain energy for life from the plant. In return, the bacteria convert nitrogen in the air into a form the plant can use.

The relationship of the bacteria to the pea plant can be classified as an example of

(1) mutualism.
(2) commensalism.
(3) parasitism.
(4) predation.
(5) saprophytism.

4. The remora, or suckerfish, attaches itself to a shark, where it receives scraps of food and protection. The shark can live either with or without the remora; it receives neither benefit nor harm from the remora.

The relationship of the shark to the remora can be classified as an example of

(1) mutualism.
(2) commensalism.
(3) parasitism.
(4) predation.
(5) saprophytism.

5. The beef obtained from cattle is an important source of protein and other nutrients in the diet of the average American.

The relationship of a person to a steer can be classified as an example of

(1) mutualism.
(2) commensalism.
(3) parasitism.
(4) predation.
(5) saprophytism.

6. The mistletoe attaches itself to a pine tree branch and extracts nutrients from it for its own use without benefiting the pine tree in return. The relationship of a mistletoe to a pine tree can be classified as an example of

(1) mutualism.
(2) commensalism.
(3) parasitism.
(4) predation.
(5) saprophytism.

7. Bread mold grows on the surface of bread. The mold uses whitish filaments to digest and absorb the nutrients in the bread. The relationship of bread mold to bread can be classified as an example of

(1) mutualism.
(2) commensalism.
(3) parasitism.
(4) predation.
(5) saprophytism.

Item 8 is based on the following table.

Characteristics of Metals

- usually solid at ordinary temperatures
- show a luster or shine
- good conductors of heat and electricity
- can be hammered into different shapes

8. You are given a small sample of an element. As you turn it over in your hands to examine it, you notice that it is yellowish in color, irregularly shaped, and hard to the touch. You finally decide that it is a nonmetal because it

(1) feels cold in your hand.
(2) feels heavy for its size.
(3) crumbles when you apply pressure.
(4) has a bright, reflective surface.
(5) has no odor.

9. A species is defined as a group of living organisms that can interbreed. Which of the following groups of organisms are not members of the same species?

(1) a Siamese cat and a house cat
(2) a German shepherd dog and a sheep dog
(3) a lion and a tiger
(4) a French man and a Spanish woman
(5) a yellow rose and a red rose

Items 10 to 13 refer to the following diagram.

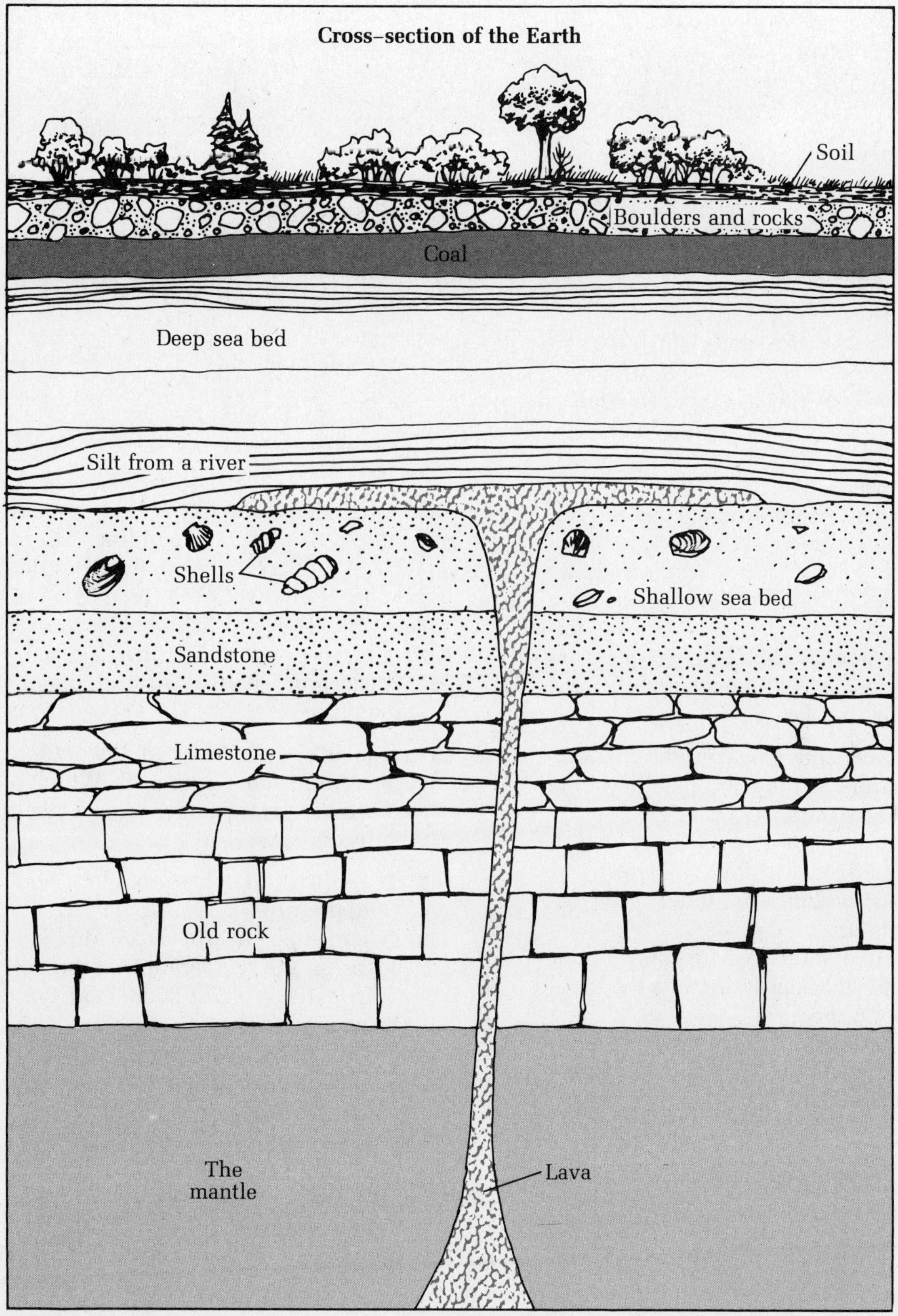

10. Based on the sketch, in which layer of Earth does lava originate?

(1) mantle
(2) limestone
(3) sandstone
(4) river silt
(5) coal bed

11. A geologist is trying to determine whether a sea-bed formation represents a deep sea bed or a shallow sea bed. Which of the following most likely would indicate a shallow sea bed?

(1) a location above sandstone
(2) the presence of shells
(3) the absence of shells
(4) a lava flow through the formation
(5) a location under a deep sea bed

12. When ocean currents carry cool water into a warm region, the change in temperature causes calcite to drop out of the ocean water. The calcite forms into limestone. Based on this information and that in the sketch, which of the following statements gives the main reason for the presence of limestone in the area shown by the sketch?

(1) The area shown was under the Arctic circle at the time the limestone formed.
(2) The area shown is a tropical beach at this time.
(3) The area shown is under an ocean at this time.
(4) The area shown was located under a tropical ocean at the time the limestone formed.
(5) The area shown was located under an ocean at the time the sandstone formed.

13. Which of the following statements supports the conclusion that the area shown in the sketch never has undergone an earthquake?

(1) The layers do not show any signs of the folding or displacement that occurs during an earthquake.
(2) The coal deposit is near the top of the layers.
(3) The lava has forced its way through several layers.
(4) The river formation is found above the shallow sea bed, indicating it replaced the sea bed.
(5) The rock record indicates that the area often was under water, although it is not at this time.

---

14. Cold-blooded animals take on the temperature of their surroundings. Cold-blooded animals include fish, reptiles, and amphibians. Warm-blooded animals, the birds and mammals, can hold their body temperatures high and constant despite changes in the outside temperature.

On a spring day in Chicago, the outdoor temperature is 60° Fahrenheit. Which of the following statements is most likely to be true?

(1) A robin is inactive with low body temperatures.
(2) A catfish in nearby Lake Michigan takes on the temperature of the air.
(3) A frog sitting on the shore has a body temperature of 60°F.
(4) A cat's body temperature is 60°F.
(5) A snake's body temperature is 80°F.

Item 15 is based on the following graph.

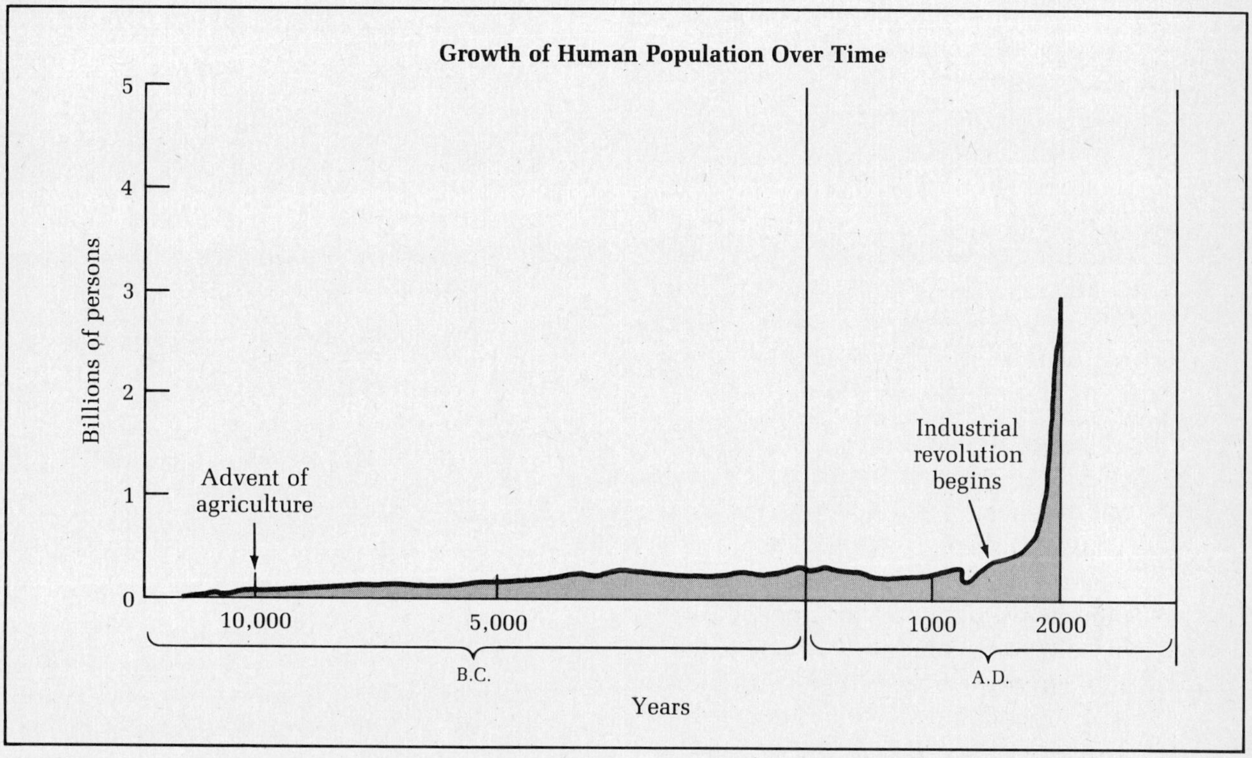

15. Which of the following is a conclusion that can be drawn from the graph?

(1) The advent of agriculture accounted for a sudden rapid increase in population.
(2) The Earth can support only about two billion people at any one time.
(3) The advent of agriculture and the industrial revolution both contributed to significant increases in population.
(4) The Industrial Revolution made the Earth less habitable, leading to a decrease in population.
(5) The growth of the human population has not been significantly affected by any external factors.

Items 16 to 19 refer to the following passage.

Sound waves are vibrations that travel by compressing and expanding matter, such as air and water. Particles of air or water move back and forth as the sound wave travels. When they move together, they are compressed. As a sound wave passes through, there is a series of compressions. The number of compressions per second arriving at a given point is the wave's frequency. The greater the frequency, the higher the pitch of the sound. The strength of the vibrations is called the wave's amplitude. The greater the amplitude, the louder the sound.

The nervous system is involved in hearing. When someone speaks to you, the sound waves are gathered by your outer ears. The outer ears direct the waves to your eardrums, causing them to vibrate. The vibrating eardrums cause three small bones in each ear to move back and forth, which in turn sets fluid in your ear moving in waves. As these waves pass over receptors inside your ear, nerve impulses are created that then travel to your brain, where they are interpreted as sound.

16. Your neighbor's stereo is much too loud. You wish your neighbor would decrease the volume by

(1) increasing the amplitude.
(2) decreasing the amplitude.
(3) decreasing the pitch.
(4) increasing the frequency.
(5) decreasing the frequency.

17. An ambulance is speeding toward you from behind. As the ambulance comes closer, the pitch of the siren rises. Which of the following statements best explains the rise in pitch?

(1) The ambulance driver increases the siren's frequency as the ambulance approaches.
(2) The ambulance "catches up" with its siren as it moves, so that you hear more wave compressions per second the closer it comes.
(3) The siren increases in amplitude as it approaches.
(4) The siren decreases in amplitude as it approaches.
(5) As the siren approaches, the increased loudness makes your ears more sensitive, so that you hear a higher pitch.

18. What effect does a loud sound have on an eardrum? A loud sound

(1) has no effect.
(2) causes the eardrum to stop vibrating.
(3) causes the eardrum to vibrate slower.
(4) causes the eardrum to vibrate harder.
(5) causes the eardrum to reflect vibratins back to the outer ear.

19. Which of the following observations provides evidence that sound waves are a form of energy?

A. A plucked guitar string vibrates back and forth.
B. The sonic boom created by a jet traveling faster than the speed of sound can break a window.
C. Concentrated sound waves are used to destroy kidney stones during ultrasound treatment.

(1) A and B only
(2) A and C only
(3) B only
(4) C only
(5) B and C only

---

20. A catalyst is a substance that increases the speed of a given chemical reaction. It does this by reducing the energy required for the reaction, allowing the reaction to proceed at a lower temperature or a faster rate.

The body uses protein catalysts to break down carbohydrates into simpler sugars. One reason it needs catalysts is that digestion takes place at body temperature and the reaction would be too slow without them.

Which of the following statements is most likely the other reason catalysts are needed?

(1) Body temperatures would become too hot without the use of catalysts.
(2) The body uses less energy during the reaction and thus has a greater net gain of energy through the use of catalysts.
(3) Catalysts speed the reaction by raising temperatures in the body.
(4) The protein in the catalysts is used in the formation of simpler sugars.
(5) The reaction would not take place without the aid of catalysts.

21. DDT poison seems to become less effective over time in controlling unwanted insects, because in fact the insects become resistant to DDT. Any population has several natural variations that arise through genetic mutation. Among insects, some usually are more resistant than others to the effect of DDT. Over time this resistant strain comes to make up a higher percentage of the population.

Which of the following is most likely the role of DDT in changing an insect population over time?

(1) DDT causes new mutations.
(2) DDT eliminates some members, leaving resistant ones.
(3) DDT preserves weak forms that come in contact with it.
(4) DDT increases the insects' rate of reproduction.
(5) DDT strengthens weak members, which become resistant.

<u>Items 22 to 26</u> refer to the following passage.

Like other characteristics, the behavior of organisms is the outcome of evolution. If an animal or plant behaves in a way that allows it to survive and to leave descendants, its genes will remain in the population's gene pool. Some of its descendants probably will behave in the same way. If its behavior causes it to die before it can leave offspring, its genes are removed from the population's gene pool.

Animals often act in ways that benefit their relatives, even at some cost to themselves. When they do so, the genes they share with their relatives are likely to remain in the population's gene pool.

Sometimes animals must use a mixed strategy of behaviors. For example, if a fiddler crab always fights other crabs, it is not likely to live long. If it never fights, it is not likely to get enough food or space. The crab's chances for a long life are best if it fights sometimes and submits at other times.

22. Acorn woodpeckers live in groups and share in the upbringing of young birds. How does this benefit the adults?

(1) The hard work of raising young is shared by others.
(2) The young can be provided with a lot of food, thus all birds eat well.
(3) The group always benefits when work is shared.
(4) The young are related to the adults.
(5) The young might be able to help the adults later.

23. Among some singing insects, males that produce the most elaborate songs are most likely to be accepted as mates by females. What is the best explanation for this?

(1) Those males have a genetic advantage over other males.
(2) Other males are incapable of reproducing.
(3) Those males will help to care for the offspring.
(4) Singing appeals to females of all species.
(5) Singing is affected by sex hormones.

24. The larvae of a certain kind of fruit fly burrow under the skin of citrus fruit, where they feed on the soft fruit. Based on this information and that in the passage, which of the following probably would be the response of a larva placed on a vertical surface near a light source? The larva would

(1) move downward and toward or away from the light.
(2) move downward and toward the light.
(3) move upward and away from the light.
(4) move downward and away from the light.
(5) move upward or downward and away from the light.

25. If a predator is active at night, which of the following survival strategies would its usual prey most likely develop?

(1) attacking the predator at night
(2) being more active at night
(3) being preyed on by a different predator
(4) being more active in the daytime
(5) being more active in winter

26. A certain animal, the prey, must feed in a meadow at least every other day to survive, although it prefers to feed every day. But 50% of the time, a predator will be in the meadow and kill any prey it finds. What is the best survival strategy for the prey group to use?

(1) Always avoid the meadow.
(2) Feed in the meadow every day.
(3) Feed in the meadow every other day.
(4) Feed in the meadow twice a day.
(5) Feed in the meadow every fourth day.

Item 27 refers to the following graph.

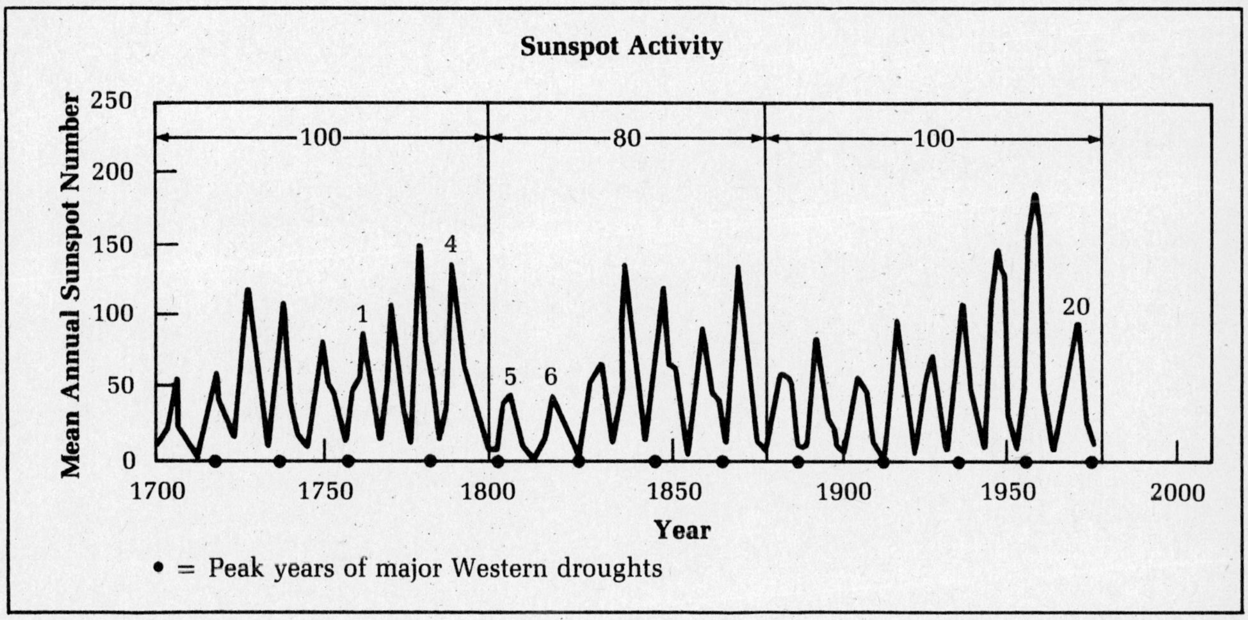

Sunspots are magnetic disturbances on the sun's surface. A greater number of sunspots usually appears when the sun is putting out greater radiation, known as a sunspot maximum.

27. Some researchers believe that sunspot activity affects the Earth's climate. They forecast that the next major drought in the West will occur around the year 2000. Which of the following conclusions drawn from the graph is the basis for this forecast?

(1) The major droughts occur with every sunspot maximum.
(2) The major droughts occur with every sunspot minimum.
(3) The major droughts occur with every other sunspot maximum.
(4) The major droughts occur with every other sunspot minimum.
(5) The major droughts occur every 30 years.

28. Electromagnetic waves can be reflected, or bounced, off a surface. Nearly all the light we see is reflected. The reason we see colors is that the chemicals in an object of that color absorb all the wavelengths of visible light except for the one of that color. That wavelength is reflected back to us.
A ripe strawberry absorbs all wavelengths of visible light except

(1) blue.
(2) green.
(3) orange.
(4) red.
(5) yellow.

Items 29 and 30 refer to the following table.

MAJOR FOOD PRESERVATIVES

Preservative	Used In
Salt (sodium chloride)	Cured meats, pickles, sauerkraut
Acids	Fruits, cured meats, pickles
Smoke	Smoked meats
Sulfur dioxide	Alcoholic beverages
Sugar	condensed milk, canned fruits, jams, and jellies

29. Your doctor recommends a diet low in sodium. Which of the following foods should you avoid?

(1) grape jelly
(2) wine
(3) canned peaches
(4) condensed milk
(5) cured ham

30. The treatment for diabetes mellitus includes eating a diet low in carbohydrates, such as sugar. Which of the following foods should a diabetic avoid?

(1) orange marmalade
(2) lunch meat
(3) pickles
(4) sauerkraut
(5) smoked ham

Items 31 and 32 refer to the following passage.

Plaque, the colorless film found on teeth, is the leading cause of dental disease in the United States. Plaque is made up of bacteria, bacterial byproducts, and saliva. The bacteria feed on sugar and starchy foods. Plaque begins to build up on teeth almost as soon as it is removed, and within a day, it hardens into a coating known as tartar or calculus. Hard-to-reach plaque and tartar can be removed only by a dentist.

The bacteria in plaque secrete acids that break down tooth enamel and irritate the gums. The gums become red and swollen and bleed easily, a condition called gingivitis. Gingivitis is treated by removing plaque and tartar from below the gum line. Untreated gingivitis develops into periodontitis. The gum tissue breaks down and pulls away from the teeth, forming pockets where more plaque and food can gather. Eventually, the jawbone becomes infected and the teeth are lost.

31. Which of the following individuals is most likely to contract gingivitis?

(1) A man who eats candy often, brushes his teeth twice a day, and visits the dentist regularly.
(2) A woman who brushes her teeth after every meal and visits the dentist regularly.
(3) A man who eats sweets and starchy foods often, brushes his teeth twice a day, and never visits the dentist.
(4) A man who had gingivitis once and now brushes and flosses his teeth daily and visits the dentist regularly.
(5) A woman who eats starchy pasta often, brushes and flosses her teeth after every meal, and visits the dentist regularly.

32. Many dentists say that there are two serious problems they tackle in their practices: 1) getting people to come in for regular cleaning of their teeth; and 2) having people do a good job of keeping their teeth clean between visits. Which of the following facts demonstrates what serious problems these are?

(1) Plaque can be found in most adult mouths.
(2) Built-up plaque hardens into tartar.
(3) Certain adults' mouths are inhospitable to teeth-damaging bacteria.
(4) Untreated gingivitis develops into periodontitis.
(5) Periodontitis is the major cause of tooth loss in adults.

33. A chemical change occurs when a substance breaks down chemically to form a new substance with new properties.

Which of the following is evidence that the digestion of protein into amino acids is a chemical change?

A. Protein and amino acids both contain nitrogen.
B. A stomach enzyme pulls apart proteins into amino acids.
C. Most proteins must be broken down into individual amino acids before they can be used by the body.

(1) A only
(2) B only
(3) C only
(4) A and C only
(5) B and C only

Check your answers on pages 129–131.

HALF-LENGTH PRACTICE TEST

Performance Analysis Chart

Directions: Circle the number of each item that you have correct on the Half-length Practice Test. Count how many items you have correct in each row; count how many items you have correct in each column. Write the amount correct per row and column as the numerator in the fraction in the appropriate "Total Correct" box. (The denominators represent the total number of items in the row or column.) Write the grand total correct over the denominator **33** at the lower right corner of the chart. (For example, if you have 29 items correct, write *29* so that the fraction reads *29*/**33**.) Item numbers in boldface represent items based on graphic material.

Item Type	*Biology (Chapter 2)*	*Earth Science (Chapter 3)*	*Chemistry (Chapter 4)*	*Physics (Chapter 5)*	TOTAL CORRECT
Comprehension	9, 14	**10, 11**		16	**/5**
Application	3, 4, 5, 6, 7		**29, 30**	1, 8, 28	**/10**
Analysis	**15,** 21, 22, 23, 24, 25, 26, 31	2, **12**	20	17, 18	**/13**
Evaluation	32	**13, 27**	33	19	**/5**
TOTAL CORRECT	**/16**	**/6**	**/4**	**/7**	**/33**

The chapters named in parentheses indicate where in the Science instruction of the Cambridge Comprehensive Program *(Unit V) and the* Cambridge Program for the Science Test *(Unit II) you can find specific instruction about each area of Science.*

On the chart, the item types are classified as Comprehension, Application, Analysis, and Evaluation. For reinforcement of the instruction for these problem types, refer to the appropriate sections of Unit III of the Comprehensive Book and Unit I of the Program For the Science Test.

FULL-LENGTH PRACTICE TEST

Directions: Choose the one best answer to each of the following questions.

1. Newton's second law of motion states, in part, that the greater the mass of an object, the harder it is to accelerate using a given amount of force. In other words, the greater the mass, the less the acceleration. Which of the following vehicles uses the most fuel to accelerate from 0 to 55 miles per hour?

(1) motorcycle
(2) motor scooter
(3) small car
(4) station wagon
(5) truck

2. Earth scientists use fossils to find out about the Earth's past. Because fossils are the remains or imprints of organisms that lived long ago, they provide clues about those organisms. Scientists also use fossils to date events in Earth's history. Which of the following is <u>not</u> a fossil?

(1) animal track in a rock
(2) volcanic rock on the seabed
(3) dinosaur bone in a riverbed
(4) petrified wood on the forest floor
(5) fish skeleton in sandstone

Items 3 to 7 are based on the following passage.

Your cells constantly are dying and being replaced by new cells. The new cells mature, then die in their turn. After a time, nearly every cell in your body has been replaced by a new one.

The deep layer of your skin makes new cells to replace dead ones. The life cycle begins as two daughter cells are born from one cell during mitosis. Each daughter cell then begins to grow. In a skin cell the first growth stage takes about 8 hours. Next, the cell duplicates its chromosomes, taking about 6 hours. This is the synthesis stage. Then the cell goes through a second growth stage for another 5 hours. Finally it is ready to divide into two cells. In the mitosis stage, which lasts about an hour, two daughter cells are produced. The two new cells are exact duplicates of their parent cell.

Not all cells go through this life cycle in the same way. Some cells, such as nerve cells, develop a specialized function instead of entering the synthesis stage. Nerve cells grow long, branchlike fibers used to transmit messages to and from the brain and spinal cord. A nerve cell will spend the rest of its life in the first growth stage. In fact, most cells in many-celled organisms, such as humans, become specialized in some function. They then stop dividing.

Cells that continually reproduce in the way described are usually in growing tissue. The cells of bone marrow, where red blood cells are produced, are one example. The cells of a plant's stem tip and root tips are another. Cancer cells also go through this life cycle. The problem with cancer cells is that they never stop dividing, so that a large number of cells build up and begin to damage the tissue around them.

3. About how long does it take a skin cell to complete its life cycle?

(1) about eight hours
(2) about one day
(3) about two days
(4) about one week
(5) about two weeks

4. A cell is duplicating its chromosomes. Which statement is true regarding this cell?

(1) The cell is a cancer cell.
(2) The cell is growing at this time.
(3) The cell is involved in sexual reproduction.
(4) The cell is preparing for mitosis.
(5) The cell is in its second growth stage.

5. The stem tip of a tomato plant is pinched off after the plant reaches half its full height. What effect will this have on the plant?

(1) The plant's upward growth will be slowed.
(2) The plant's root development will be slowed.
(3) The plant's sideways development will be slowed.
(4) The plant will produce fewer tomatoes.
(5) The plant will die.

6. Which of the following microscope slides most likely would show each stage of the cell's life cycle? A slide showing cells from

(1) an onion root tip.
(2) a frog's nerve.
(3) a maple leaf.
(4) a cat brain.
(5) a petunia petal.

7. What is the major advantage to the body in having an old, dying cell produce daughter cells exactly like itself?

(1) Such daughter cells are easy for the body to make.
(2) The body uses less energy to make such daughter cells.
(3) Such daughter cells are best suited to take on the parent cell's function.
(4) Such daughter cells live longer than their parent did.
(5) Such daughter cells tend to become specialized in function.

Item 8 refers to the following chart.

Oral Drug Pathway in the Body

Mouth
↓
Stomach
(acid)
↓
intestines (alkaline) → excretion from body
↓
blood → excretion from body
↓
target organ

8. A drug for treating kidney disease is harmful to the stomach but still needs to be broken down before it can enter the blood stream. The drug is best coated with a substance that causes the drug to

(1) break down in the stomach
(2) resist the stomach's acid but break down in the intestine
(3) resist the stomach's acid and the intestine's alkalinity
(4) be excreted by the intestine
(5) be excreted by the blood

Item 9 is based on the following information.

The heart muscle receives oxygenated blood through the coronary artery, which branches to various parts of the heart muscle. Some families are genetically disposed to deposit more fatty substances in their arteries. The arteries then are narrowed. High blood pressure and obesity also can contribute to narrowing. If the coronary artery is narrowed by fatty deposits, too little oxygen reaches the heart muscle. Angina, a warning pain, might occur. A heart attack might follow, damaging the heart muscle. When the muscle is damaged by lack of oxygen, it releases two enzymes (call them A and B) not ordinarily found in the blood.

9. A patient arrives at the hospital complaining of chest pain. A diagnosis of heart attack is most likely if

(1) the patient is very obese and has high blood pressure.
(2) the coronary artery is found to be very narrow.
(3) the patient has a family history of heart attacks.
(4) a diagnosis of angina is made.
(5) enzymes A and B are found in the patient's blood.

Items 10 to 13 refer to the following graph.

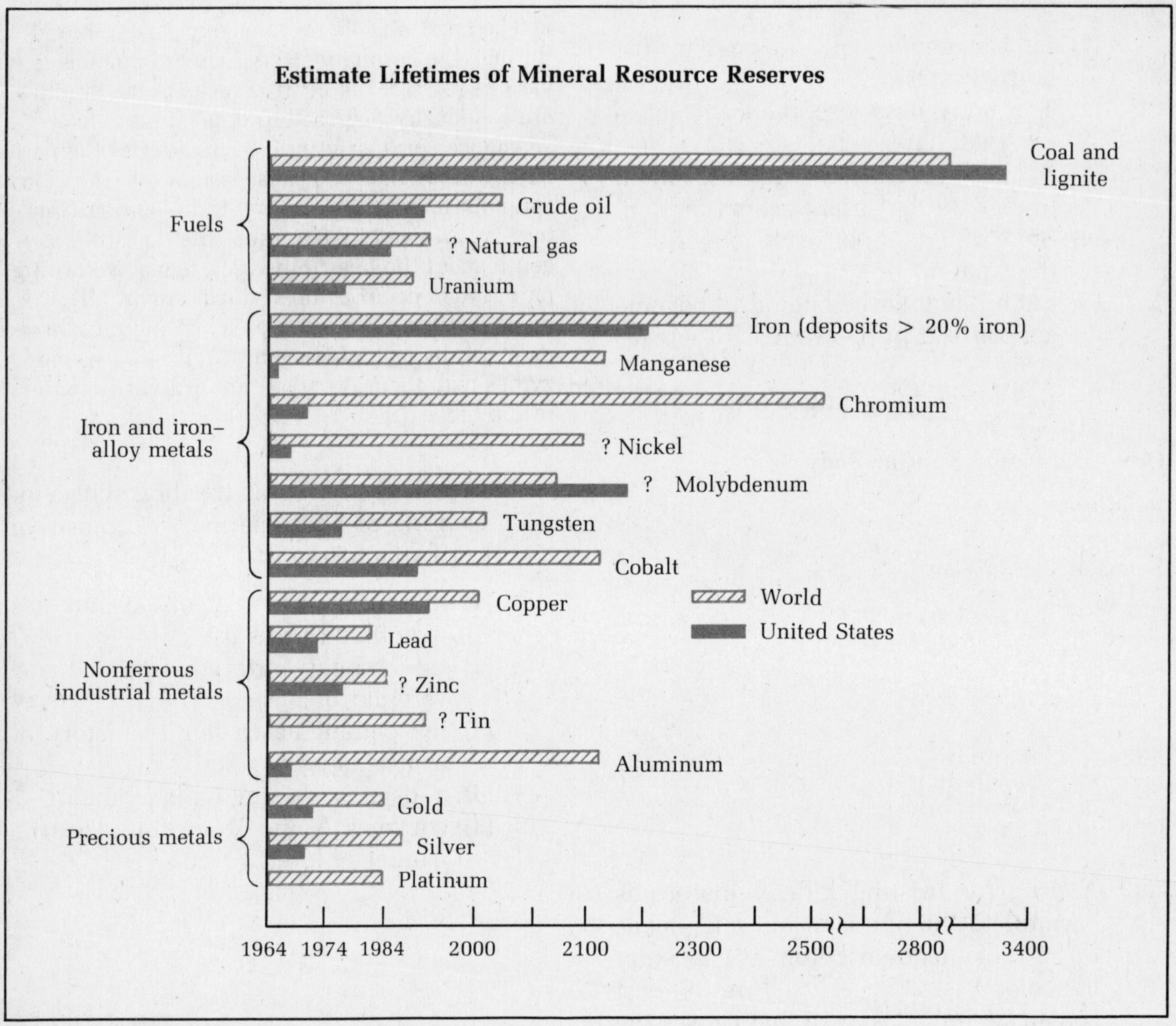

10. Which of the following metals is expected to last the longest?

(1) chromium
(2) cobalt
(3) copper
(4) gold
(5) silver

11. An American manufacturer needs a metal that is in plentiful supply in the United States. Because of the manufacturing process, an iron or iron-alloy metal cannot be used. Which of the following metals would best suit the manufacturer's needs?

(1) aluminum
(2) copper
(3) lead
(4) platinum
(5) silver

12. After the oil shortages of the 1970s, caused in part by foreign oil producers, many utilities and other industries in the United States switched from burning oil to burning coal for energy. Which of the following was probably industry's main consideration in the switch?

(1) Burning coal causes less pollution than burning oil.
(2) Coal is easier than oil to extract from the ground.
(3) Coal is easier than oil to transport from place to place.
(4) Coal contains more energy than the same amount of oil.
(5) Coal is more abundant than oil in the United States.

13. A certain geologist warns that <u>all</u> of Earth's mineral resources will be depleted in 200 years unless present patterns of use are changed. The lifetime(s) of which of the following minerals on the graph refute this belief?

A. coal
B. chromium
C. lead
D. aluminum

(1) A and B only
(2) A and C only
(3) B and C only
(4) B and D only
(5) C and D only

14. A healthy woman jumped rope for 5 minutes. Her muscle cells used oxygen to burn sugar molecules for energy. This process, called respiration, is demonstrated by the fact that during the 5 minutes, her

(1) blood pressure increased from 115/75 to 130/75
(2) heart rate increased from 70 to 140 beats per minute
(3) blood oxygen level decreased from 100 to 92 units
(4) temperature increased from 98.6° to 98.9° F.
(5) body lost 2 cubic centimeters of water as sweat

Item 15 refers to the following diagram.

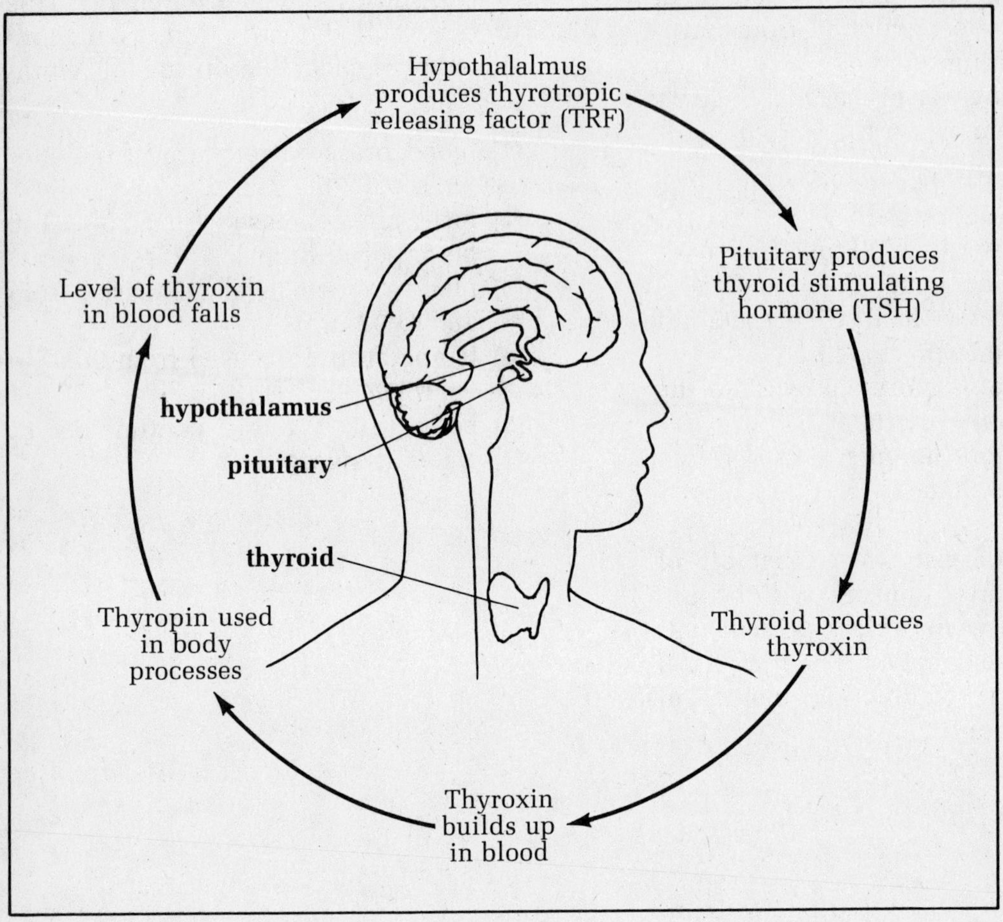

15. An injury to the pituitary gland causes lowered production of TSH. What is likely to happen?

(1) less TRF produced by the thyroid
(2) less thyroxin produced by the hypothalamus
(3) more TRF available for body processes
(4) more TSH produced by the hypothalamus
(5) less thyroxin available for body processes

Items 16 to 19 refer to the following passage.

A fluid is any substance that can flow. A fluid can be either a liquid or a gas. There are certain facts known about fluids. One is that fluids exert pressure in every direction. In a fluid at rest, this pressure increases with the depth of the fluid.

A second fact about fluids is known as Pascal's principle: If you apply pressure anywhere to a confined fluid, this added pressure is transmitted to every part of the fluid and its container. This means that if you apply inward pressure on a piston of small area, it results in a large push outward on a large piston. The pressure is the same on both pistons, but the area of the second piston is larger, and thus the total force on the second piston is greater as well.

A third fact is known as Archimedes' principle: A fluid exerts an upward push on an object totally or partly submerged in it. This upward push is called a buoyant force. The object appears to lose an amount of weight equal to the weight of the displaced fluid.

Two facts about moving fluids often are used by engineers. The first is that in a tube, where the tube area is large, the fluid speed is small, and vice versa. The second is called Bernoulli's principle: The pressure in a fluid decreases as the fluid speeds up, and vice versa.

16. Which fact about fluids explains the increased pressure on his body a scuba diver notices as he descends to the ocean floor?

(1) The pressure of a fluid increases with depth.
(2) Pascal's principle
(3) Archimedes' principle
(4) The speed of a fluid depends on the tube area.
(5) Bernoulli's principle

17. An airplane wing is flat underneath and curved on top to give it lift. Air rushing over the curved top of a wing travels farther and thus faster than the air flowing under the wing. The pressure above the wing decreases, and the airplane lifts off because of which fact?

(1) The pressure of a fluid increases with depth.
(2) Pascal's principle
(3) Archimedes' principle
(4) The speed of a fluid depends on the tube area.
(5) Bernoulli's principle

18. When you push on a car's brake pedal, the pedal pushes on a piston in the master cylinder. The increase in pressure is transmitted through the brake fluid to each wheel cylinder. This is an application of which fact?

(1) The pressure of a fluid increases with depth.
(2) Pascal's principle
(3) Archimedes' principle
(4) The speed of a fluid depends on the tube area.
(5) Bernoulli's principle

19. In a river in which the width remains the same throughout, the water flows swiftly where the river is shallow, but slowly where the river is deep. This is explained by which fact?

(1) the pressure of a fluid increases with depth.
(2) Pascal's principle
(3) Archimedes' principle
(4) The speed of a fluid depends on the tube area.
(5) Bernoulli's principle

20. Injured plant tissue gives off large quantities of ethylene gas. Healthy plant tissue that comes into contact with even a small amount of ethylene begins to produce the gas itself. Ethylene gas causes fruit tissue to ripen and then rot. Which of the following strategies should an apple farmer use to avoid damage to his or her crop?

(1) Leave the damaged apples to help ripen those that are still green.
(2) Pick through apples daily to remove any damaged fruit.
(3) Remove damaged apples only when there are many that are damaged.
(4) Remove damaged apples from the storage area weekly.
(5) Pick apples before they ripen, then store with no further care.

21. The body has a variety of defenses against disease. Unbroken skin and the mucous membrane in the mouth serve as barriers to disease-causing microorganisms. However, microorganisms can enter through the eyes and nose. Stomach acid killls many of the microorganisms in food. Most of the microorganisms that do manage to infect the body are destroyed or neutralized by the immune system.

Which of the following situations is most likely to cause you to catch a cold?

(1) You get caught in a rainstorm and get soaked through your clothing.
(2) You pick up a cold virus on your hands and then rub your eyes.
(3) You eat food contaminated by a cold virus.
(4) A friend with a cold kisses you on the cheek.
(5) You work near a person with a cold who is coughing a lot.

Items 22 to 26 refer to the following passage.

Plants are an essential part of life on Earth. Animals depend on plants for two things basic to life: food and oxygen. During photosynthesis, plants make a simple carbohydrate called glucose. They do this by combining parts of two molecules, carbon dioxide and water, in the presence of sunlight. Carbon dioxide and water molecules contain oxygen. Some of the oxygen is used in making glucose. The rest is given off by the plant. Plants also make another food—proteins—from nitrogen in the environment.

There are many types of plants. Some, such as mosses, have no real roots or stems to bring water to their cells. As a result, every part of these plants must be able to absorb water directly from the environment. Other species, such as flowering plants, have a complex structure for transporting food and water.

The roots of a plant have three main functions. The first is to anchor the plant in the soil. The second is to absorb water and other nutrients, substances that supply nourishment to the plant. The third function of roots is to store food for the plant. The stem supports the plant and transports nutrients between the roots and leaves. The leaves are where most plants make food. Carbon dioxide enters the leaf through openings in its surface. Most leaves are flat to expose a greater area to the sunlight needed for photosynthesis.

Flowers, fruits, and seeds are the reproductive organs of a plant. First the flower grows. After pollen joins with the egg inside the flower, the petals fall off. The inside of the flower grows larger to become the fruit. The fruit stores food for the seed. Fruits also help scatter seeds by rolling along the ground or being eaten by animals. When the seed finds a favorable environment, it sprouts and grows into a new plant.

22. Which of the following equations represents photosynthesis?

(1) carbon dioxide + water $\xrightarrow{\text{sunlight}}$ oxygen + glucose
(2) oxygen + water $\xrightarrow{\text{sunlight}}$ carbon dioxide + glucose
(3) carbon dioxide $\xrightarrow{\text{sunlight}}$ oxygen + water + glucose
(4) carbon dioxide + oxygen $\xrightarrow{\text{sunlight}}$ water + glucose
(5) oxygen + water $\xrightarrow{\text{glucose}}$ sunlight + carbon dioxide

23. A utility company is building solar collectors in the Arizona desert. Based on similar structures in a plant, which surface would collect the most sunlight?

(1) a shallow bowl, curving away from the sunlight
(2) a deep bowl, curving away from the sunlight
(3) a shallow bowl, curving toward the sunlight
(4) a deep bowl, curving toward the sunlight
(5) a flat surface

24. If an impenetrable coating were sprayed on a plant's leaves, what process would this interfere with?

(1) the absorption of water
(2) the absorption of minerals
(3) the transportation of water
(4) the transportation of nutrients
(5) photosynthesis

25. Which of the following observations supports the statement that every part of a moss must absorb water directly from the environment in order to live?

(1) Mosses live in shady, moist areas.
(2) Mosses need water to reproduce.
(3) Mosses tend to be short so that water transportation is unnecessary.
(4) If a moss is left half in water and half out, the dry side will die.
(5) Mosses are made of spongy tissue that absorbs water easily.

26. To test whether or not a plant needs soil to grow, a scientist takes two identical plants and puts one plant in soil and another in tap water. The containers for each plant are of equal size. The plants then are placed side by side and given equal amounts of water and sunlight. The plant in soil grows larger than the plant in water. The scientist concludes that a plant needs soil to grow.

Which of the following factors could have been the variable that produced the experimental results?

(1) sunlight.
(2) water.
(3) plant type.
(4) container size.
(5) availability of nutrients.

Item 27 refers to the following diagram.

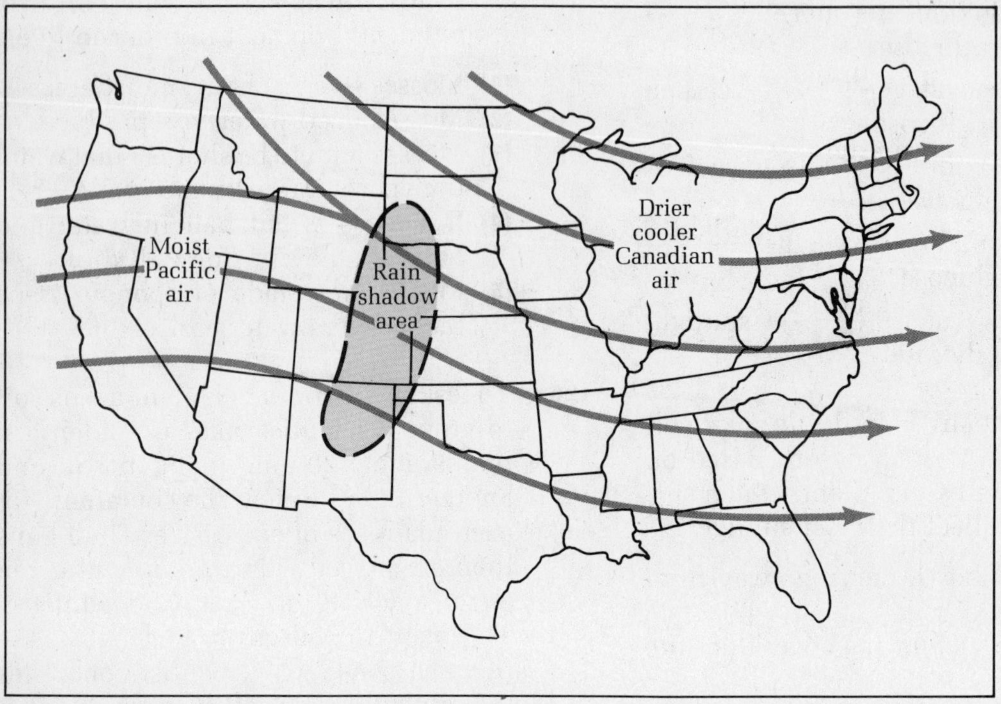

The diagram shows the average pattern of wind flow across the United States, projected for the next 20 years.

27. Based on the weather map, 5 weather forecasters made the predictions listed below concerning weather patterns in the United States for the next 20 years. Which of the predictions is an incorrect interpretation of the map's data?

(1) The Southwest will be under the influence of moist Pacific air for much of the time projected.
(2) It will be consistently drier in the Northeast than in the Southeast.
(3) The area of the country in the rain shadow area will tend to be drier than most other areas.
(4) Most of the northern states can expect cool summers and winters for the next 20 years.
(5) The states of the far West will continue to have warm, moist summers and winters.

28. Cohesion is a force that holds water molecules together. Water molecules tend to do this best where they meet a different medium. Water molecules on the waxed surface of a car hood bead up as they meet the wax and air. A tiny water strider can walk on the "skin" formed by the water molecules on a pond's surface. Even though metal is denser than water, a carefully placed razor blade floats on the water in a cup. When dropped instead, the razor sinks. Why does it sink?

(1) The air above the water is disturbed.
(2) The blade molecules are lighter.
(3) The impact breaks forces between molecules.
(4) The water molecules change in density.
(5) The water molecules are broken into atoms.

Items 29 and 30 are based on the following diagram.

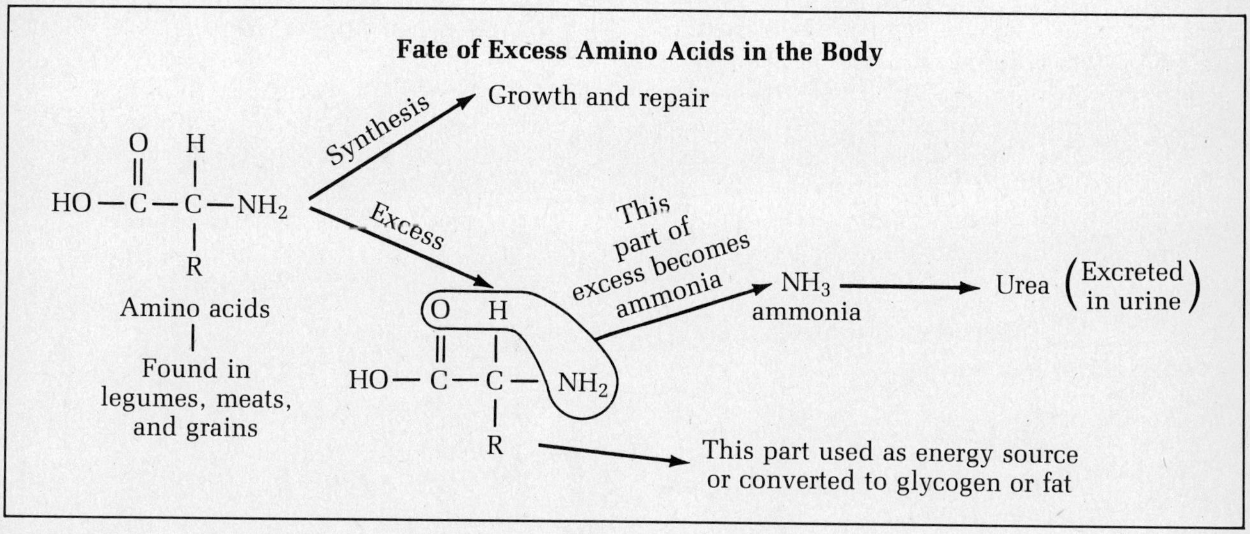

29. Which of the following could happen to a person who often eats more meat than his or her body requires?

A. increased urination
B. weight gain
C. increased cellular growth and repair

(1) A only
(2) B only
(3) C only
(4) A and B only
(5) B and C only

30. Which statement is evidence that excess amino acids cannot be stored as is by the body?

(1) Excess amino acids are used for growth and repair.
(2) Excess amino acids are stored as fat.
(3) Amino acids contain several different elements.
(4) Excess amino acids are broken down into other components.
(5) Ammonia is converted to urea and excreted in urine.

Items 31 and 32 refer to the following passage.

Most scientists believe that life arose from nonliving matter billions of years ago. As part of the effort to prove this theory, a scientist filled a container with a mixture of gases like those thought to have existed when life first appeared on Earth. He then sent electrical sparks through the container. After a time, amino acids, the building blocks of protein, formed.

31. What did the scientist assume in carrying out the experiment?

(1) Amino acids are the building blocks of protein.
(2) Amino acids are found in living organisms.
(3) The formation of simple proteins was the next step after amino acid formation.
(4) Amino acids rarely form from gases.
(5) An electrical energy source, such as lightning, played a part in creating life.

32. Which of the following experiments would support the belief that the forming of protein from amino acids led to the development of the first one-celled organisms?

(1) creating one-celled organisms from a mixture of gases
(2) proving one-celled organisms contain protein
(3) creating one-celled organisms from various proteins
(4) developing larger organisms from one-celled organisms
(5) forming protein from amino acids

---

33. Which of the following statements made by the makers of a popular brand of orange juice could be proven by chemical analysis of the juice?

A. Contains no preservatives or artificial flavors.
B. Citrus Garden juice is a good source of Vitamin C.
C. The flavor is light and tangy.
D. Contains only the finest Florida oranges.

(1) A only
(2) B only
(3) A and B only
(4) B and C only
(5) C and D only

34. Particle size influences the rate at which a given material dissolves. This is because smaller particles have a greater surface area per given amount of material than do larger particles. The greater the surface area, the faster a material dissolves. Assuming all other factors are equal, which of the following drugs will dissolve and enter the bloodstream the fastest?

(1) acetaminophen caplet (Tylenol)
(2) aspirin
(3) powdered pain reliever (Alka-Seltzer)
(4) multivitamin tablet
(5) ibuprofen capsule (Advil)

Items 35 to 39 are based on the following information.

Sedimentary particles are classified according to size. The fine particles tend to stick together, whereas the coarse particles do not. The finer the particles, the less drainage of liquids they provide. Below are five particle types classified from coarse to fine.

		DIAMETER
Coarse	(1) cobbles:	from 64 to 256 millimeters (size ranges from one to several tennis balls)
	(2) pebbles:	from 2 to 64 millimeters (size ranges from a bead to a marble)
	(3) sand:	from 1/16 to 2 millimeters (size ranges from a pen point to a match head)
Fine	(4) silt:	from 1/256 to 1/16 millimeter (about the size of a pin point)
	(5) clay:	less than 1/256 millimeter (invisible to the unaided eye)

Each of the above categories might be used more than once in the following set of items.

35. A contractor wants to install tubing to drain the water under a football field. The guidelines for installing the perforated tubing state that in choosing the bedding material that will surround the tube, the contractor should avoid fine particles that could enter and block the line. He also should avoid large rocks that could damage the tubing. Which choice of bedding would best suit the contractor's needs?

(1) cobbles
(2) pebbles
(3) sand
(4) silt
(5) clay

36. A house builder wants to install tubing to channel water from a downspout to a nearby storm sewer. The installation guidelines call for small, loose particles of soil that will flow around the tubing and minimize soil settling. Which of the following particles should the builder use?

(1) cobbles
(2) pebbles
(3) sand
(4) silt
(5) clay

37. In choosing a covering for a footpath, you wish to use a material that will provide good drainage and not stick to the shoes when wet. The material should be large enough to provide a firm surface, yet not so large as to be difficult to walk on. Which of the following particles would be most appropriate?

(1) cobbles
(2) pebbles
(3) sand
(4) silt
(5) clay

38. Moving water tends to separate fine particles from coarser particles. The coarse particles then settle to the bottom. In a cup of water from a river, you observe fine particles suspended in the water. Which of the following particles are you seeing?

(1) cobbles
(2) pebbles
(3) sand
(4) silt
(5) clay

39. Oil usually is found under a fine-grained rock known as marine shale. A geologist searching for oil would have the most success drilling through rock that was at one time either

(1) cobbles or pebbles.
(2) pebbles or sand.
(3) sand or silt.
(4) silt or clay.
(5) clay or cobbles.

Item 40 is based on the following sketch of the number of nerve impulses triggered by warm and hot stimuli respectively.

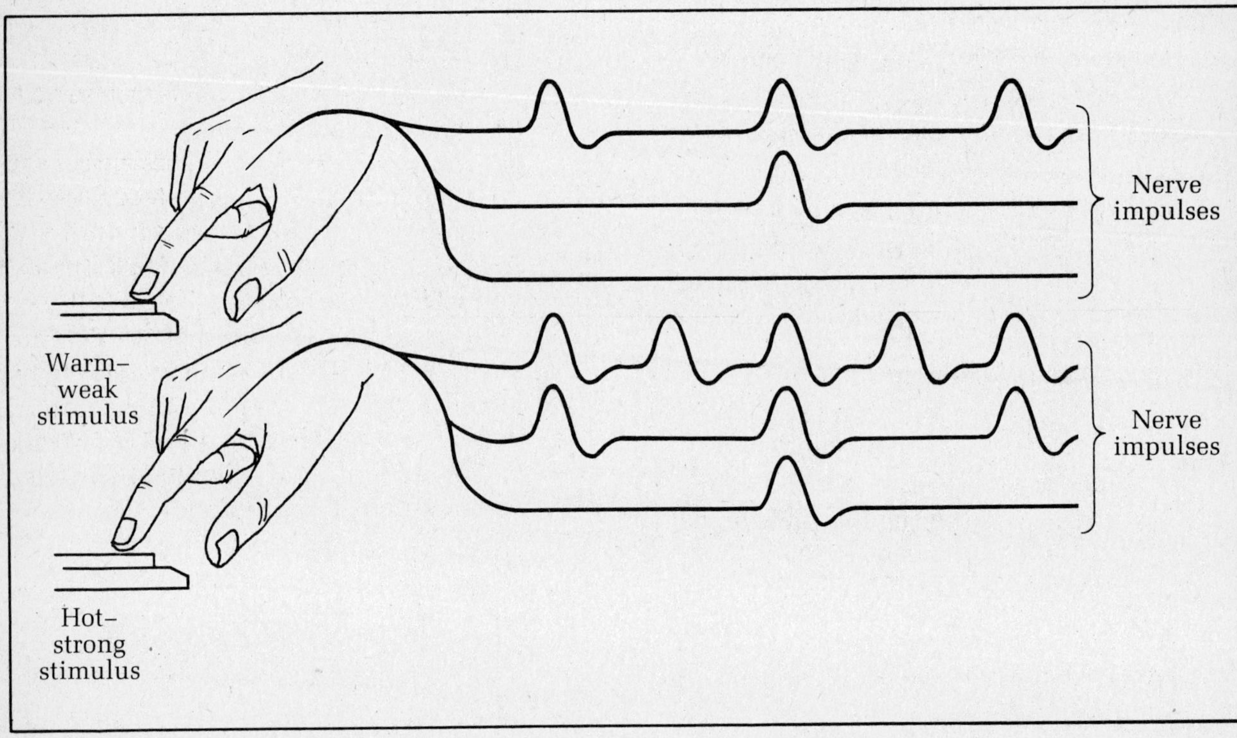

40. Which of the following factors is (are) most likely used by the brain in determining whether it is receiving nerve impulses from a strong or weak stimulus?

A. kind of nerve impulses received
B. number of nerve impulses received
C. number of neurons transmitting

(1) A only
(2) B only
(3) C only
(4) A and B only
(5) B and C only

---

41. Elm trees once lined the streets of many U.S. cities. But in the 1930s a fungus was brought in on logs from Europe. The fungus was carried by bark beetles that live in elms. Millions of elms died from the Dutch elm disease, leaving whole streets bare of trees. What measure can be taken to prevent the spread of such disease in the future?

(1) Find a disease-resistant substitute for the elm and plant it widely.
(2) Plant a variety of healthy trees in every area.
(3) Use stronger fungicides and insecticides.
(4) Import trees from Asia instead of Europe.
(5) Find a cure for Dutch elm disease.

<u>Items 42 to 46</u> are based on the following diagram.

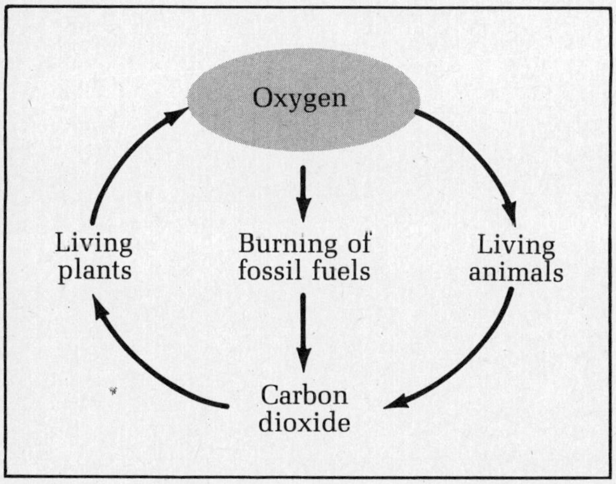

42. Which statement best summarizes the relationship between plants and animals?

(1) Plants and animals help each other by each giving off gases into the air that the other one needs.
(2) Plants help animals by giving off oxygen, but animals do not help plants.
(3) Animals help plants by giving off carbon dioxide, but plants do not help animals.
(4) The burning of fossil fuels takes away all the oxygen plants give off, thus animals do not benefit from plants.
(5) Plants and animals both benefit from the burning of fossil fuels.

43. In a closed aquarium, the animals and plants are balanced as to production and use of oxygen and carbon dioxide. If more fish were added, which of the following events would occur?

(1) The amount of carbon dioxide produced by the plants would decrease.
(2) The amount of oxygen in the water would increase.
(3) The amount of oxygen produced by the plants would decrease.
(4) The amount of carbon dioxide in the water would increase.
(5) The amount of carbon dioxide in the water would decrease.

44. As the human population of the world increases, more fossil fuels and forests are being used for human purposes. Based on the diagram, what effect(s) are these actions having on the atmosphere?

(1) Forest removal is raising the level of oxygen in the atmosphere; fuel burning is lowering it.
(2) Forest removal is lowering the level of oxygen in the atmosphere; fuel burning is raising it.
(3) Forest removal is raising the level of oxygen in the atmosphere; fuel burning also is raising it.
(4) Neither forest removal nor fuel burning is raising the level of carbon dioxide in the atmosphere.
(5) Both forest removal and fuel burning are raising the level of carbon dioxide in the atmosphere.

45. If the burning of fossil fuels were removed from the cycle in the diagram, what effect would this have?

(1) There would be more oxygen and less carbon dioxide.
(2) There would be more carbon dioxide and less oxygen.
(3) Animals would die.
(4) Plants would die.
(5) All life as we know it would die.

46. It sometimes is argued that pollution from burning fossil fuels eventually will damage the delicate balance of nature. How does the diagram support this?

(1) There will be more plants, making more fossil fuels to burn, thereby causing more pollution.
(2) Burning fossil fuels increases levels of carbon dioxide and decreases levels of oxygen; plant and animal life will suffer.
(3) Burning fossil fuels will cause animals to need more carbon dioxide and plants to need more oxygen.
(4) Animals will be eliminated from the cycle and only plants will survive.
(5) Plants will be eliminated from the cycle and only animals will survive.

47. Atmospheric pressure often is measured by the height of a thin tube of mercury in a barometer. The atmosphere pushes down on the surface around the tube, and the mercury is pushed up into the tube. It is measured at sea level and at 0° C because elevation and temperature both affect air density. Gravity pulls the molecules in air toward the Earth. Thus, air is slightly less dense and weighs less at the top of a mountain than at sea level. Standard atmospheric pressure is 760 millimeters (mm) of mercury (1 inch = 2.54 mm).

A normal barometric pressure in millimeters at a mountaintop 5,000 feet high is

(1) 12 mm.
(2) 54 mm.
(3) 200 mm.
(4) 754 mm.
(5) 760 mm.

Item 48 refers to the following diagram.

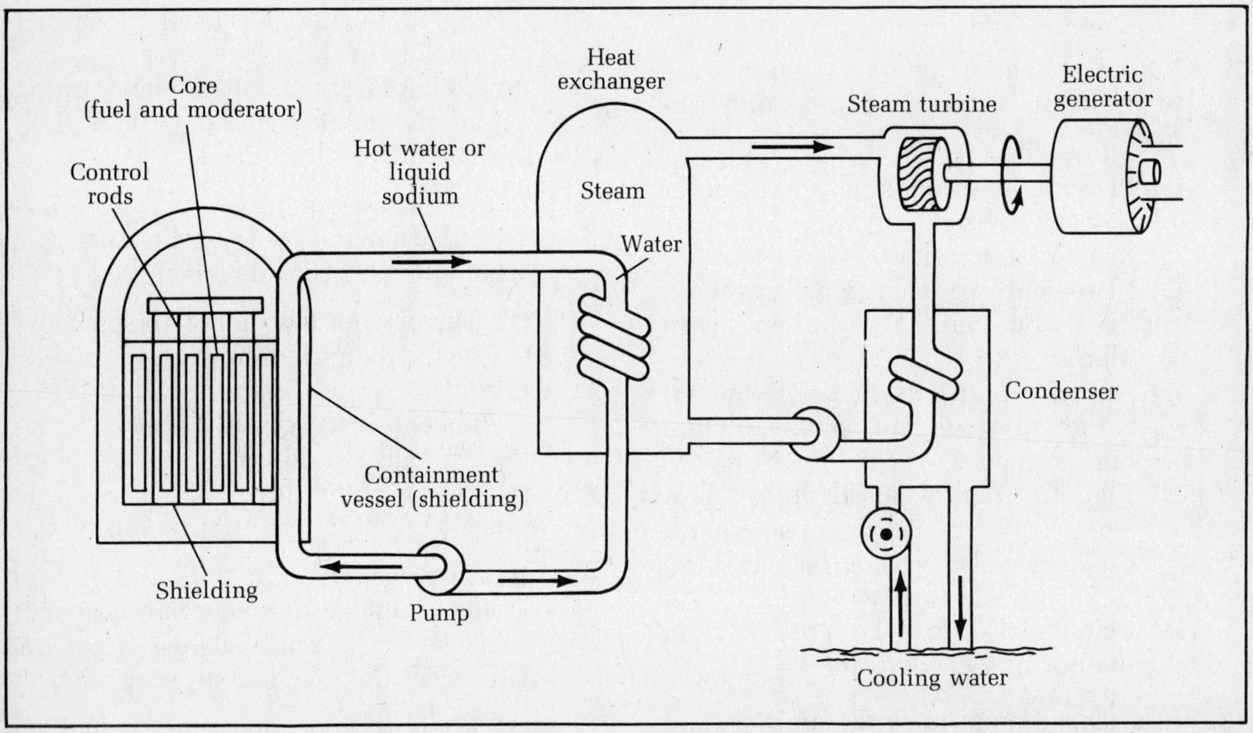

48. Which of the following forms of energy is not used at some point in a nuclear reactor?

(1) nuclear
(2) heat
(3) steam
(4) motion
(5) electric

Items 49 to 53 are based on the following passage.

Whales and walruses affect the Bering Sea ecosystem not only by what they eat but also by digging in the sea floor for food. Gray whales excavate huge pits as they search for their favorite food, a shrimplike animal called an amphipod. The whale sucks in sediment, then filters it through its baleen, straining the food out of the mud. The comblike baleen captures the prey, which is swallowed by the whale. The whale then spits the mud out of its mouth into the water.

This disturbance of the sea floor helps make the area a productive one, ensuring the whales' food supply. After the whales expel mud into the water, currents carry the mud particles away. The sand left behind is ideal for amphipods to grow in. The whales' actions also release previously buried nutrients into the water, feeding plankton, which in turn are eaten by the amphipods. Finally, the whales' baleen allows young, small amphipods to escape being eaten by slipping through the baleen's openings. The young amphipods are left behind to recolonize the sand. The amphipods burrow into the sand, where they construct the tubes in which they live. The tubes hold the sand in place, helping to slow erosion by water currents. The young amphipods then grow until the whales return a year later.

The Pacific walruses that live in the Bering Sea also affect the sea floor. Walruses dig narrow furrows in the sea floor as they search for clams and other prey. However, these narrow furrows probably don't affect the cycle of nutrients as deeply as do the whale pits.

49. What relationships would be most similar to the relationship between whales and young amphipods?

(1) parents and children
(2) cows and horses
(3) wolves and lambs
(4) plows and crops
(5) sunshine and plants

50. Which of the following human activities would the whales' feeding habits interfere with most?

(1) shipping on the Bering Sea
(2) laying oil pipelines on the Bering Sea floor
(3) scuba diving in the Bering Sea
(4) fishing in the Bering Sea
(5) collecting Bering Sea animals for a zoo

51. What effect would mining the Bering Sea for sand have on the whale population? The whale population probably would

(1) decrease.
(2) increase slightly.
(3) stay the same.
(4) increase for a time, then decrease.
(5) increase greatly.

52. The sand on the Bering Sea floor was deposited during an Ice Age and will not be replaced until another glacier covers the area. Which of the following actions would cause an immediate increase in erosion of the sand layer?

(1) poisoning the amphipods through an oil spill
(2) hunting the whales
(3) hunting the walruses
(4) fertilizing the plankton
(5) feeding the amphipods

53. Which of the following statements would an environmental preservation group, such as Greenpeace, be least likely to use in support of leaving the whales and walruses undisturbed by human activities?

(1) Whales benefit the Bering Sea ecosystem.
(2) About 200,000 walruses make their home in the Bering Sea.
(3) Walruses eat clams that could be harvested by fishermen.
(4) Amphipods are the whales' preferred food during the summer.
(5) An oil spill could lead to the whales' starvation.

54. Living organisms can grow, reproduce, move, use energy, and respond to their environment. Which of the following statements supports the view that a virus is a living organism?

(1) A virus remains one size from its beginning to its end.
(2) A virus does not use energy.
(3) A virus can use a cell to produce more viruses.
(4) A virus does not respond to outside conditions.
(5) A virus is spread through the movements of living organisms.

55. Coal is formed over millions of years as plants die, are covered up by other plants, and eventually are compressed and heated into coal. Most coal formed from plants that lived between 200 million and 400 million years ago, when much of the Earth's land was covered with mosses and ferns. Which of the following statements best explains why coal is considered a nonrenewable natural resource?

(1) mosses and ferns no longer exist on Earth.
(2) It takes so long for coal to form that by the time new coal appears, the people now living will be dead.
(3) Because so many ecosystems have been destroyed, there are not enough plants left to form coal.
(4) The plants that form the best coal are extinct.
(5) Plant litter is raked and cleaned away before it can be buried.

<u>Items 56 to 59</u> refer to the following passage.

Polymers are among the most important chemical compounds to the consumer. Polymers are large molecules made up of chains of repeating chemical units. The chains often contain thousands of such units.

Most of the products of the petrochemical industry are polymers. These include plastic wrap, nylon, rayon, and rubber. Polymers usually are strong, light, flexible, and long-lasting. Some resist conduction of heat and electricity.

Polymers, such as plastics, have replaced glass, metal, and paper in many household products. Toothpaste tubes and milk bottles are made of plastic polymers. Polymers also are found in the home as polyester sheets and clothes, vinyl floors, and nylon-blend rugs.

56. Based on the passage, which of the following is <u>not</u> a petrochemical product?

(1) plastic bag
(2) glass baby bottle
(3) phonograph record
(4) credit card
(5) nylon-bristled tooth brush

57. A chemical engineer is looking for a way to use a new heat-resistant plastic he has developed. Which of the following items would make the best use of his invention?

(1) toy
(2) rug
(3) curtain
(4) screwdriver handle
(5) food cooler

58. Which of the following would benefit <u>least</u> from the properties of plastic?

(1) eyeglass frames
(2) ink pen
(3) car interiors
(4) table top
(5) typewriter casing

59. Some people argue that the petrochemical industry is hazardous to human health. Which of the following statements supports this viewpoint?

(1) Plastic automobile engines are more fuel-efficient than metal engines.
(2) Plastic is used to make artificial limbs and joints for people.
(3) Workers in polyvinyl chloride (PVC) factories have a higher chance of getting liver disease.
(4) Soft contact lenses cause little eye irritation because they allow water to pass through them.
(5) Plastic plates and cups usually do not break when dropped.

Item 60 refers to the following figure.

Stand Replacement Cycle in a Ponderosa Pine Forest

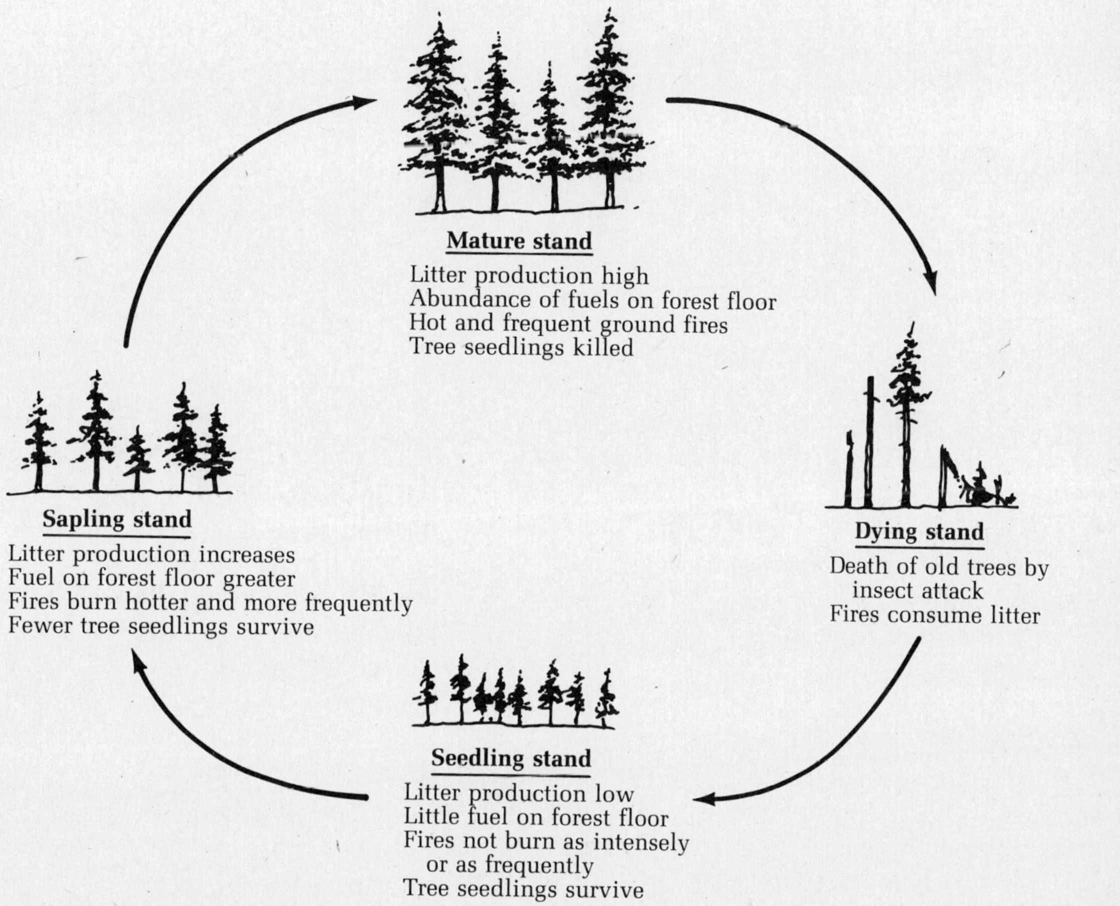

60. A forest resource management center allows controlled burning of forests. Using the information from the diagram on natural stand replacement, what is one important reason for allowing controlled forest fires?

(1) Controlled fires are a safe way to burn old trees and make space for tree seedlings.
(2) Controlled fires are not dangerous to forest-dwelling animals.
(3) Increasing numbers of people are building homes in forests.
(4) Controlled fires burn forest litter, making it easier for people to walk in the forest.
(5) Controlled fires allow old trees to continue maturing without competition from the seedlings.

61. Platelets, part of the body's blood clotting machinery, show an increased tendency to stick together at certain times of day. A common time for platelets to stick together is in the hours just after a person gets out of bed after a long sleep. If a clump of platelets gets stuck in an oxygen-supplying artery, the blockage can cause a heart attack.

Which of the following generalizations about the likelihood of heart attacks does this information support?

(1) People with coronary artery disease are more likely to have heart attacks.
(2) Smokers are more likely to have heart attacks.
(3) People with a family history of heart disease are more likely to have heart attacks.
(4) Heart attacks are most likely to occur just after an afternoon nap.
(5) Heart attacks are most likely to occur in the morning after sleeping all night.

Items 62 to 64 are based on the following sketch.

Capillary Action

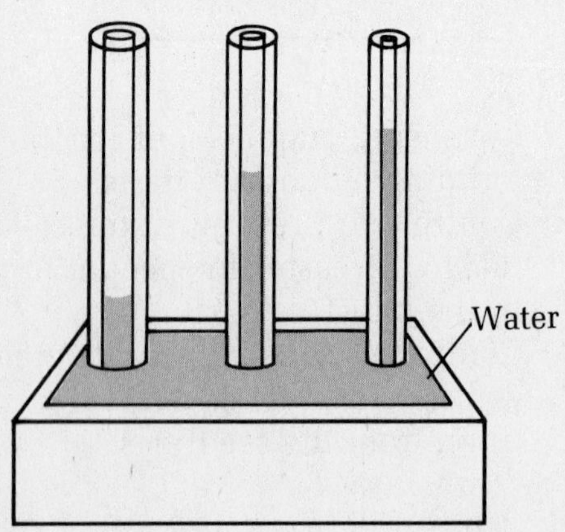

62. Which of the following statements is true of capillary action?

(1) The water rises highest in the widest tube.
(2) The water rises highest in the narrowest tube.
(3) The water's height is unaffected by the different tubes' diameters.
(4) The water is pushed upward by gravity.
(5) The water is pushed upward by centrifugal force.

63. Plant xylem is similar to a narrow tube. Water moves through xylem partly because of capillary action. Using the information provided, it can be reasoned that the xylem transports water from

(1) the roots to the leaves.
(2) the leaves to the roots.
(3) the leaves to the stem.
(4) the leaves to the flowers.
(5) the seeds to the roots.

64. What effect would placing more water in the container have on the amount of water in the tubes?

(1) The water in the widest tube only would rise.
(2) The water in the middle tube only would rise.
(3) The water in the narrowest tube only would rise.
(4) The water in all three tubes would rise.
(5) There would be no change in any of the tubes.

65. Your bones serve several functions in your body. They are levers that produce movement when the muscles attached to them contract. They support and protect body structures. The bones also store minerals, such as calcium and phosphorus, which give them their hardness and heaviness.

Suppose a pregnant woman does not take in enough calcium to meet her baby's needs. Her body must supply the remainder. Which of the following could happen? Her bones could

(1) cease acting as levers.
(2) become lighter and softer.
(3) cease supporting the body.
(4) cease protecting the heart and brain.
(5) become detached from the muscles.

66. Some studies have shown that children belonging to minority races have low average IQ scores. It is difficult to control these studies for environmental effects, such as poverty and cultural differences. What generalization regarding IQ tests can be drawn from this?

(1) IQ scores are not an indication of intelligence.
(2) IQ scores cannot be used accurately to compare people from different backgrounds.
(3) Intelligence is a genetic characteristic.
(4) Hunger and stress are causes of low IQ scores.
(5) Intelligence differs from person to person.

Check your answers on pages 131–133.

FULL-LENGTH PRACTICE TEST
Performance Analysis Chart

Directions: Circle the number of each item that you have correct on the Full-length Practice Test. Count how many items you have correct in each row; count how many items you have correct in each column. Write the amount correct per row and column as the numerator in the fraction in the appropriate "Total Correct" box. (The denominators represent the total number of items in the row or column.) Write the grand total correct over the denominator **66** at the lower right corner of the chart. (For example, if you have 60 items correct, write *60* so that the fraction reads *60*/**66.**) Item numbers in boldface represent items based on graphic material.

Item Type	*Biology (Chapter 2)*	*Earth Science (Chapter 3)*	*Chemistry (Chapter 4)*	*Physics (Chapter 5)*	TOTAL CORRECT
Comprehension	3, 4, 9, 14, 22, **42,** 65	2, **10**	56	**48, 62**	**/12**
Application	6, **15,** 21, 23, **29, 43,** 49, **60**	**11, 35, 36, 37, 38, 39,** 47	**8,** 20, 34	1, 16, 17, 18, 19	**/23**
Analysis	5, 7, 24, 26, 31, **40,** 41, **44, 45,** 50, 51, 52	**12,** 55	57, 58	28, **63, 64**	**/19**
Evaluation	25, **30,** 32, **46,** 53, 54, 61, 66	**13, 27**	33, 59		**/12**
TOTAL CORRECT	**/35**	**/13**	**/8**	**/10**	**/66**

The chapters named in parentheses indicate where in the Science instruction of the Cambridge Comprehensive Program *(Unit V)* and the Cambridge Program for the Science Test *(Unit II) you can find specific instruction about each area of Science.*

On the chart, the item types are classified as Comprehension, Application, Analysis, and Evaluation. For reinforcement of the instruction for these problem types, refer to the appropriate sections of Unit III of the Comprehensive Book and Unit I of the Program For the Science Test.

SIMULATED TEST

Directions: Choose the one best answer to each of the following questions.
Time: 95 minutes

1. Water molecules evaporate, or change into a gas, when they reach a high enough velocity to escape liquid form. The higher the temperature, the more molecules have enough energy to evaporate.

Your city has put a water conservation measure into effect. You may water your lawn or garden only 15 minutes per day. Which of the following times of day normally would cause the least evaporation and most efficient use of water?

(1) late morning
(2) noon
(3) early afternoon
(4) late afternoon
(5) early evening

2. A blizzard is a combination of high winds and falling snow that lasts more than a few hours. Because the wind is so strong, the temperature feels much colder than it really is. As a result, any skin that is exposed to the wind and snow can quickly become frost-bitten. The blowing snow can reduce visibility to zero, and snowdrifts pile up rapidly. Which of the following would be the worst action to take if trapped in a car during a blizzard?

(1) Listen to the radio for weather bulletins.
(2) Leave the car to find help.
(3) Yell out the window for help.
(4) Run the engine occasionally for heat.
(5) Do exercises to keep warm.

Items 3 to 7 are based on the following information.

Every organism must interact with its environment and other organisms to survive and reproduce. An animal's response to its environment is called behavior. Below are five types of animal behaviors.

A. communication:	An animal passes information on to others.
B. conditioning:	An animal learns to link one experience with another.
C. cooperation:	An animal acts in a way to benefit its group.
D. investigation:	An animal explores its surroundings.
E. trial-and-error learning:	An animal does the same task over and over until it is done correctly.

Each of the above categories may be used more than once in the following set of items.

3. A teacher discusses a new lesson with her class. This is an example of

(1) communication.
(2) conditioning.
(3) cooperation.
(4) investigation.
(5) trial-and-error learning.

4. A dog is taken to a park it never has been in. The dog trots from place to place, sniffing the air. The dog is involved in

(1) communication.
(2) conditioning.
(3) cooperation.
(4) investigation.
(5) trial-and-error learning.

5. A chicken pecks at two different levers, one red and one blue. When it pecks the red lever, nothing happens. When it pecks the blue lever, a piece of food falls into a hopper in front of the chicken. Soon it pecks at only the blue lever. Pecking the blue lever is an example of

(1) communication.
(2) conditioning.
(3) cooperation.
(4) investigation.
(5) trial-and-error learning.

6. A male bird brings food to the female while she is nesting. This is an example of

(1) communication.
(2) conditioning.
(3) cooperation.
(4) investigation.
(5) trial-and-error learning.

7. A child learning to play baseball swings at the ball again and again until he learns to hit the ball with the bat. This is an example of

(1) communication.
(2) conditioning.
(3) cooperation.
(4) investigation.
(5) trial-and-error learning.

Item 8 refers to the following chemical equation and information.

$$\underset{\text{(baking soda)}}{2NaHCO_3} + \underset{\text{(hydrogen)}}{H_2} \longrightarrow$$

$$\underset{\text{(sodium)}}{2Na} + 2H_2CO_3 \text{ (unstable)}$$

$$\downarrow$$

$$\underset{\text{(carbon dioxide)}}{2CO_2} + \underset{\text{(water)}}{2H_2O}$$

Baking soda, a leavening agent, acts in breads and cakes according to the above reaction. The expansion of carbon dioxide gas and water vapor when heated in the oven leads to the rising of the dough.

8. Increasing the baking soda in a bread recipe would result in which of the following products?

(1) a thin, crackerlike bread
(2) a fluffy, airy bread
(3) a heavy, compressed bread
(4) a soft-crusted bread
(5) a hard-crusted bread

9. In designing a pond to mimic the workings of the ocean, a biologist chose several species of fish. The first was mirror carp, which swim so strongly they churn and mix the water, much as ocean currents do. The second was tilapia, which filter the water as they eat algae, much as whales filter the water for plankton. The third was white amur, which feeds on grass clippings and flower stalks, fertilizing the water with plant nutrients. In the ocean this is done by nutrient-laden currents.

The biologist chose these particular species of fish in order to

(1) stock the pond with popular game fish.
(2) study a typical pond ecosystem.
(3) keep the pond clean, healthy, and fertile.
(4) learn why the fish vary from each other.
(5) turn the pond into a miniature ocean.

Items 10 to 13 refer to the following sketch.

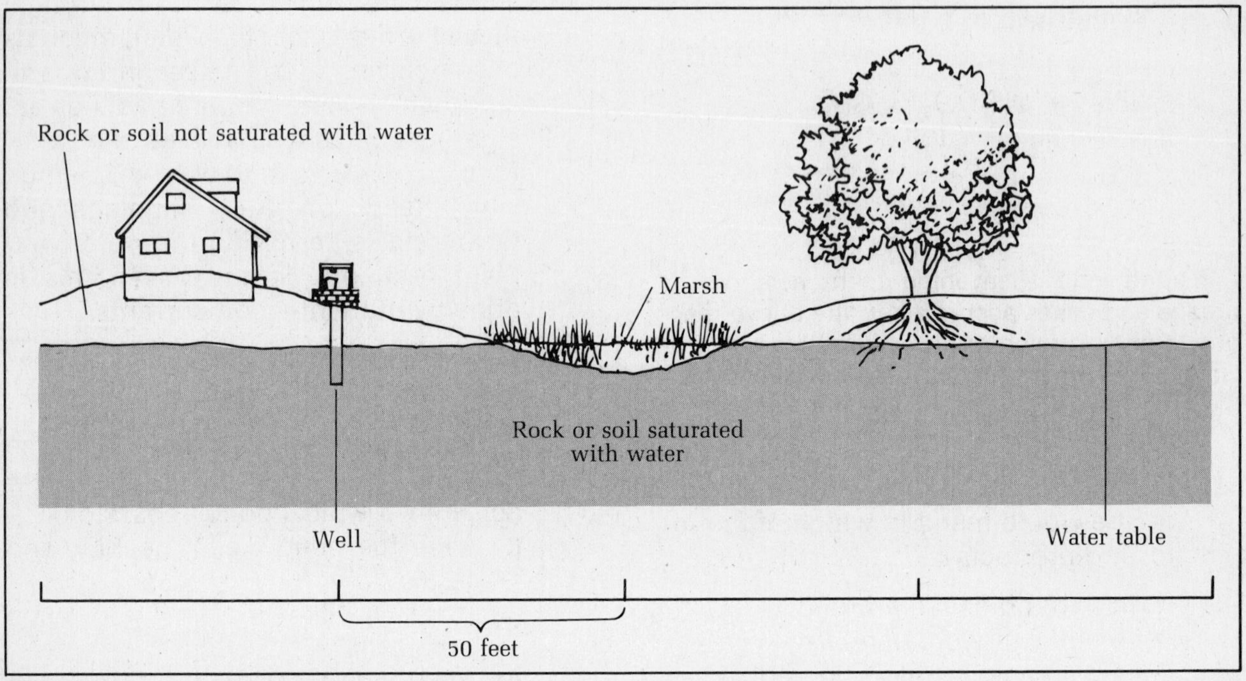

10. Which of the following is the best definition of "water table"?

(1) place where water seeps above ground
(2) any body of water
(3) zone where rock and soil are saturated with water
(4) lower boundary of the water saturation zone
(5) upper boundary of the water saturation zone

11. A builder wishes to construct a house that will not be subject to water seepage in the basement, even during rainy periods. Which of the following locations should the builder choose?

(1) a drained marsh
(2) a valley
(3) a flat piece of land
(4) the top of a hill
(5) the side of a hill

12. Which of the following would be caused by draining the marsh of all water?

A. The well would run dry.
B. The tree's roots would need to go deeper to reach water.
C. The house's basement would become flooded.

(1) A only
(2) B only
(3) C only
(4) A and B only
(5) B and C only

13. A groundwater supply can become polluted by seepage from a septic tank. To avoid contamination by sewage, a well should be at least 45 feet from a septic tank. On the other hand, the septic tank should be located fairly close to the house, no farther than 90 feet away. Based on this information, which of the following locations would be most suitable for a septic tank in the set-up pictured?

(1) to the left of the house
(2) between the house and the well
(3) between the well and the marsh
(4) between the marsh and the tree
(5) to the right of the tree

---

14. Practicing modern medicine often means applying science to the problems of the body. A physician must know how chemicals affect the body's cells in order to know when to prescribe a certain drug. Antibiotics, for example, work either by killing bacteria directly or by preventing bacteria from increasing in the body until the body's own defenses can take over. The prescription of drugs requires a knowledge of which two branches of science?

(1) biology and chemistry
(2) chemistry and physics
(3) biology and physics
(4) biology and earth science
(5) chemistry and earth science

Item 15 refers to the following chart.

The Inheritance of Color Blindness

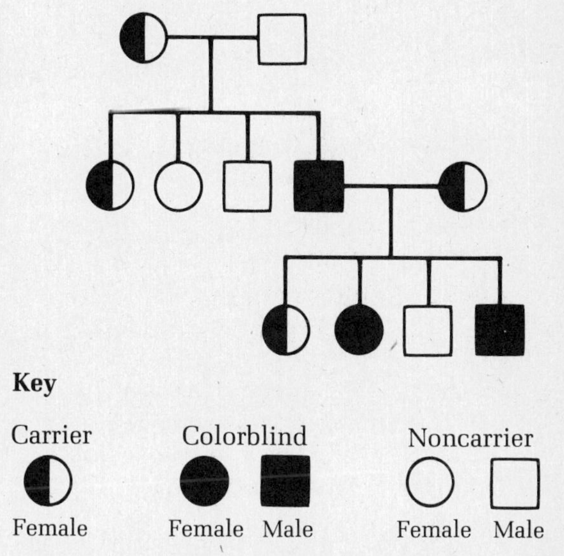

15. A normal man with a mother who carries the gene for color blindness marries a normal woman. What are the chances their children will be color blind?

(1) 0
(2) 1 in 4
(3) 2 in 4
(4) 3 in 4
(5) 4 in 4

Items 16 to 19 refer to the following passage.

Microwave ovens cook food quickly because the water in the food rapidly absorbs the energy of the waves. Microwaves are similar to light waves in that both are waves of electric fields moving to and fro. The electric field of the microwave gives water molecules more energy so that they strike one another with more vigor, raising the temperature of the water in the food.

In a conventional oven, food is heated from the outside in. The surface of food like beef becomes very hot, turning it brown and changing its flavor. The inside never becomes as hot as the outside.

When beef is cooked in a microwave oven, water is heated throughout the meat. The water then heats the solid parts. The beef's surface never becomes hotter than the boiling point of water and so does not turn dark brown.

Liquid water is heated so quickly in a microwave that, in a closed container, an escape route must be provided for the steam. Plastic cooking bags should be slit and lids of covered dishes should not fit tightly. Otherwise the steam could exert onough pressure on the container to explode it.

Metal containers should not be used in a microwave oven because metal reflects microwaves. The food is shielded from the microwave energy and will not get hot. Also, the reflected microwaves could overload the microwave emitter, damaging the oven.

16. Why doesn't meat "brown" in a microwave oven?

(1) The fats in the meat do not reach the required temperature.
(2) The containers used do not allow meat to brown.
(3) The meat cooks too quickly to brown.
(4) The meat's surface does not reach the required temperature.
(5) The water in the meat absorbs the brown color.

17. Which one of the following containers could be used safely in a microwave oven?

(1) a sealed plastic container
(2) an iron pot
(3) an open soup can
(4) a tightly closed glass jar
(5) a china bowl

18. Why is it unwise to hard-boil an egg in a microwave oven?

(1) The egg will not get hot enough, making it unsafe to eat.
(2) Steam could build up in the unbroken egg, causing it to explode.
(3) The egg yolk will change color and appear unappetizing.
(4) The white surface of the egg could reflect microwaves, causing the emitter to overload.
(5) The egg could overheat, affecting its nutrient value.

19. You wish to prepare the following foods so that they taste and appear their best. Keeping in mind how microwaves cook food, which one of the following would be most appropriate to prepare in a microwave oven?

(1) a hamburger
(2) fried potatoes
(3) a reheated roast
(4) broiled fish
(5) cookies

20. Ascorbic acid, or vitamin C, helps slow the wilting and discoloring of produce. Ascorbic acid is found in citrus fruits, broccoli, spinach, strawberries, potatoes, and cabbage. A cook is preparing a fruit salad. Which of the following will keep the fruit salad looking fresh the longest?

(1) tossing the salad with lemon juice
(2) including strawberries in the salad
(3) including apples in the salad
(4) tossing the salad with apple juice
(5) mixing mayonnaise into the salad

21. Plants can be grown in barren soil by working organic matter—such as peat, compost, or rotted manure—into the soil. In addition to fertilizing the soil, the porous organic matter allows plant roots to receive more air. Organic matter loosens heavy soil, such as clay, and improves the water-holding capacity of sandy soil.

Two new homeowners are trying to grow grass on a path that formerly was heavily traveled. Besides adding organic matter, how might they improve the soil so that the grass will grow?

A. Till the soil.
B. Compact the soil.
C. Spread fertilizer on the area.

(1) A only
(2) B only
(3) C only
(4) A and C only
(5) B and C only

Items 22 to 26 refer to the following passage.

Scientific advances in agriculture have led to a dramatic increase in food production. These advances have come in three main steps. The first was the change from animal to machine power, which greatly increased the amount of work a farmer could do. The second step was the introduction of chemical fertilizers and pesticides.

The third step involved the application of what we know about plant genetics. Plant breeders have developed crops that grow faster, produce more food per plant, and have better natural defenses against diseases and insects. Through genetic engineering, scientists can produce better plants. For example, researchers found bacterial genes that produce a protein that kills insects. They inserted these genes into tomato plant cells. The result is a tomato plant with leaves that are poisonous to insects.

Scientists also cross-breed plants to produce new varieties with desired traits. For example, varieties of corn developed to resist cool weather and mature three weeks earlier have allowed farmers in North America to grow corn 250 miles farther north than they could ten years ago.

The resulting worldwide food surplus has its negative side. Most of the modern varieties of grain grow best with chemicals, such as pesticides. Environmental experts believe the increased use of chemicals around the world is polluting water and causing illness. Farm crops also usually require more water than does the natural vegetation. In addition, as more land is used for farming, soil erosion is becoming more of a problem.

22. Which of the following is an example of cross-breeding plants?

(1) fertilizing corn so that it produces a better crop
(2) using a combine to harvest wheat more quickly
(3) inserting a wild rice gene into a domestic rice plant cell
(4) planting winter wheat in a field that normally is used for corn
(5) producing a new wheat plant from an American and a Japanese wheat plant

23. Corn breeders treat cells from immature corn kernels with toxic agents, such as herbicides. If a cell mutates to become resistant to the herbicide, it ultimately can be grown into a new plant that will produce more seeds with resistance to the herbicide. The new plant has been produced by using knowledge of

(1) plant breeding.
(2) fertilizers.
(3) genetic engineering.
(4) soil management practices.
(5) crop rotation.

24. Which of the following is the best reason why farming increases soil erosion in some areas?

(1) The natural vegetation is better suited than the farm crops to holding soil in place.
(2) The farm crops require the use of fertilizers and pesticides.
(3) The farm crops have a shorter growing season than the natural vegetation.
(4) The farm crops use more water than does the natural vegetation.
(5) The farm crops have been produced through genetic engineering and cross-breeding.

25. What is the main advantage of genetic engineering of farm crops?

(1) Some crops have been made disease-resistant.
(2) Tomato plants have been developed that have leaves that are poisonous to insects.
(3) A strain of corn has been developed that can resist cooler weather.
(4) Scientists can develop strains of plants that are hardier, more nutritious, better tasting, and more fruitful.
(5) Crops can be grown farther north than they could before.

26. Some people argue that scientific tampering with farm crops will do more harm than good. Which of the following statements supports this argument?

(1) Scientists might not be able to achieve the results they are seeking.
(2) New strains of most plants require more pesticides because they have less natural resistance to diseases and insects.
(3) Human beings should not tamper with nature.
(4) There is already a problem with distributing surplus food where it is needed; most surplus food sits and rots.
(5) The increased need for farmland would take away land that could be used for housing.

Item 27 is based on the following sketch.

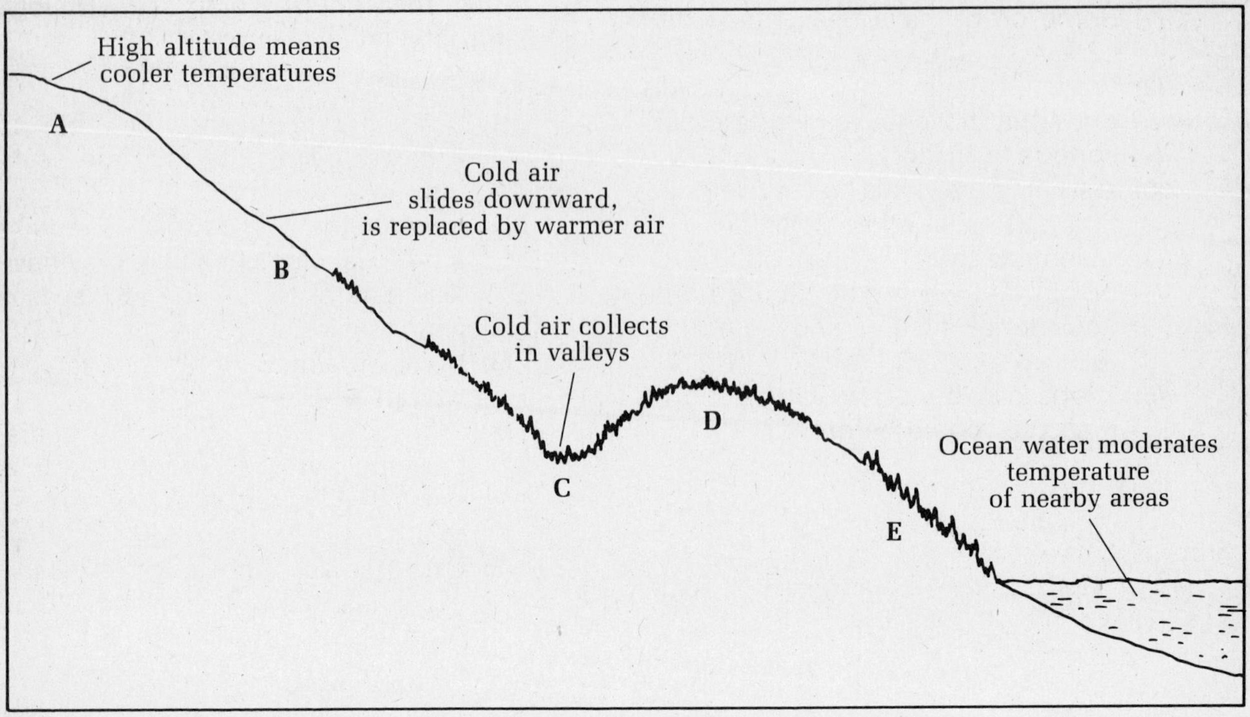

27. A well-placed garden can have a growing season up to one month longer than one in a cooler, frost-prone site. What makes site E on the diagram the best location for a garden?

(1) The garden is watered, even during droughts, by the ocean.
(2) The cold air sliding down the hill is warm by the time it reaches the garden.
(3) The slope of site E is less extreme than that of site B.
(4) The angle of the sun is better at site E than at other sites.
(5) The low altitude and nearness to the ocean make site E warmer than the other sites.

28. A group of engineers is searching for ways to cut down the friction between a car's wheels and the road, thus increasing gas mileage. The friction between the tires and the road's surface slows cars down. Below are three suggestions for reducing friction. Which most likely would reduce friction between the tires and the road?

A. Use magnetic fields to suspend the car in air an inch above the road.
B. Use six tires instead of the usual four tires.
C. Lubricate the wheel axle and other car parts frequently.

(1) A only
(2) B only
(3) C only
(4) A and B only
(5) A and C only

Items 29 and 30 are based on the following graph.

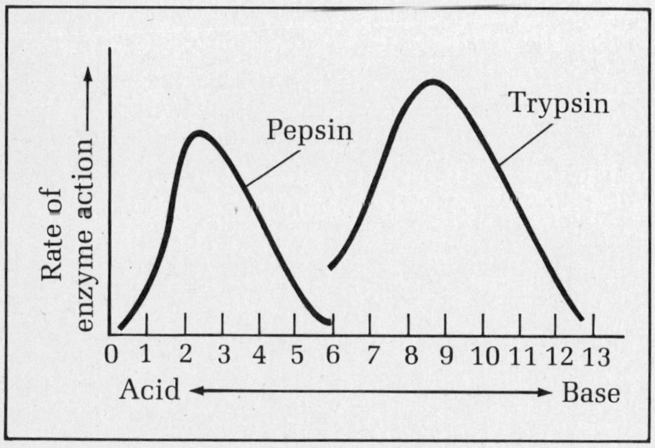

The enzyme pepsin starts the digestion of proteins in the stomach. The enzyme trypsin continues the digestion of proteins in the small intestine.

29. The pH of the small intestine probably is between

(1) 2 and 3
(2) 3 and 4
(3) 5 and 6
(4) 7 and 8
(5) 8 and 9

30. Which of the following is a hypothesis that would explain why pepsin is more effective in the stomach than trypsin?

(1) An enzyme's rate of action depends on the pH of its environment.
(2) There is more pepsin than trypsin in the stomach's juices.
(3) The stomach's contents are too acidic for trypsin to function.
(4) Trypsin is unable to start the digestion of proteins.
(5) Proteins in the stomach are too tightly bound together for trypsin to digest.

Items 31 and 32 are based on the following passage.

Most heart attacks happen when a coronary artery becomes blocked by a blood clot. Researchers designed the following experiment to find a treatment that would both clear the clot and stop a new one from forming. One group of patients was given a clot-dissolving drug only. The other group took the drug combined with coronary angioplasty. In coronary angioplasty, a tiny, balloon-tipped tube is threaded into an artery. The balloon then is inflated to push the artery open.

After treatment, the patients who received only the drug had arteries that were about 75% narrower than normal. The patients getting both tretments had only half that narrowing and also showed less new clot formation.

31. Which of the following must be assumed if the experiment's results are to be considered valid?

(1) All the patients had had previous heart attacks.
(2) The patients in both groups were matched as to age and other variables.
(3) Those patients with jobs continued working during treatment.
(4) The patients were all men.
(5) All the patients were age 55 or older.

32. Which of the following conclusions is supported by the experiment's results?

(1) People who have had one heart attack are likely to have another.
(2) Coronary angioplasty can prevent strokes as well as heart attacks.
(3) A program of exercise prevents some heart attacks.
(4) Coronary angioplasty helps prevent heart attacks.
(5) Clot-dissolving drugs can reduce high blood pressure.

---

33. Metals tend to be dense, are able to be formed into shapes, and conduct heat and electricity well.

You wish to determine whether a certain piece of pipe is made of iron or of some carbon-based plastic. Which of the following would be evidence for iron?

A. The pipe feels heavy.
B. The pipe has a curved shape.
C. When one end of the pipe touches a flame, the other end feels hot.

(1) A only
(2) B only
(3) A and B only
(4) B and C only
(5) C only

34. A cup of salt is stirred into a pot of soup until the salt is completely dissolved in the soup. The soup then is placed in the refrigerator. The next day, large crystals of salt are at the bottom of the pot. Which of the following statements best explains this crystal formation?

(1) Salt comes out of solution at lower temperatures.
(2) Salt undergoes a chemical change in the hot soup.
(3) Water pressure pushed the salt to the bottom of the pot.
(4) The salt particles grew overnight.
(5) The soup evaporated in the refrigerator.

Items 35 to 39 refer to the following passage.

Carbon dioxide plays an important part in regulating temperatures at the Earth's surface. Along with water vapor, CO_2 allows light energy to pass through the atmosphere but traps the heat energy radiating back off the Earth's surface. Because a greenhouse works in a similar way, this is known as the greenhouse effect.

Carbon dioxide gets into the air in several ways. Animals and plants release it as a product of cellular respiration. Bacteria and fungi cause its release as they break down dead plant and animal matter. Carbon dioxide also is released by the burning of some five billion tons of fossil fuels each year.

At this time there are about six tons of carbon as CO_2 over each acre of the earth's surface. Much of this carbon is taken from the air as plants use it during photosynthesis. Later it might be deposited as peat, coal, oil, or limestone. The ocean also absorbs large amounts.

What is worrisome is that more carbon dioxide is being released into the air than can be removed through natural processes. As a result, temperatures are rising a little each year. Scientists fear that this will upset the Earth's ecological balance. Desert areas could become wetter, and other areas could become drier. The polar ice caps would begin to melt, causing the sea level to rise, wiping out coastal areas. Plant growth would increase as plants took in increased levels of carbon. All these effects—some good, some bad—could be upon us long before we know how to control them.

35. Why are scientists so concerned about the greenhouse effect?

(1) Carbon dioxide is a poisonous gas.
(2) Plants cannot use all the carbon dioxide being formed.
(3) Industry must continue burning fossil fuels.
(4) The oceans will become even bigger.
(5) Life as we know it might change or even end.

36. How would cutting down an acre of trees affect the carbon cycle in that area?

(1) less carbon removed from the air; more carbon returned to the air
(2) less carbon removed from the air; less returned to the air
(3) more carbon removed from the air; less returned to the air
(4) more carbon removed from the air; more returned to the air
(5) more carbon removed from the air; the same amount as before returned to the air

37. How could erosion contribute to the increase of carbon in the air?

A. Erosion breaks down carbon-containing matter, causing its release into the atmosphere.
B. Eroded soil fills in small ponds, decreasing the amount of water available to absorb carbon.
C. Erosion carries fertile soil away to the oceans, interfering with plant growth.

(1) A only
(2) B only
(3) C only
(4) A and B only
(5) B and C only

38. Although increased levels of carbon dioxide generally would be undesirable, what would be one beneficial side-effect of increased levels of carbon in the air?

(1) increased availability of fish and seafoods
(2) better air quality
(3) increased farm productivity
(4) increased supply of fresh water
(5) increased supply of fossil fuels

39. Which of the following is evidence that there has been an increase in the greenhouse effect in the past century?

(1) Fossil fuel use has increased in the last 100 years.
(2) Frozen ground beneath the Arctic has warmed several degrees in the last 100 years.
(3) The harvesting of trees has increased in the last 100 years.
(4) Temperatures in the U.S. have cooled a few degrees over the last 100 years.
(5) Scientists believe the greenhouse effect has increased in the last 100 years.

Item 40 is based on the following figure.

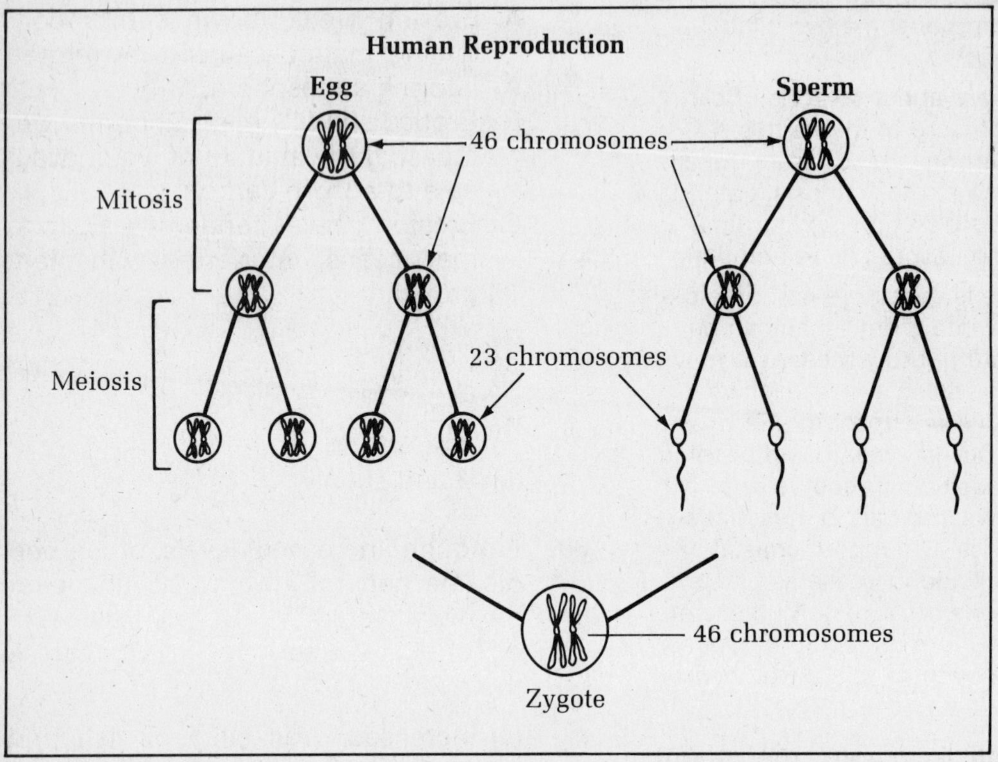

40. What would happen if meiosis did not occur during reproduction?

(1) The number of chromosomes would decrease with each generation.
(2) Only females would survive until birth.
(3) Sexual reproduction would be replaced by asexual reproduction.
(4) The number of chromosomes would double with each generation.
(5) The zygote would be twice as large.

41. Researchers have found that breast-fed babies use their mouths differently than do bottle-fed infants. Breast-feeding is more work for the mouth muscles. Also, breast-fed babies do not develop the forward tongue-thrust that bottle-fed infants do to stop too much milk from coming into their mouths. The same study found that children who were breast-fed for more than a year developed 40% fewer crooked teeth than those breast-fed for less than a year or bottle-fed.

Which of the following statements is a conclusion that could be drawn from the study's data?

(1) Sucking a finger is one cause of crooked teeth.
(2) Breast-fed infants work their mouth muscles more than do bottle-fed babies.
(3) Breast-feeding is more popular in the U.S. than it used to be.
(4) Bottle-fed infants must thrust their tongues forward.
(5) The different mouth action required for breast-feeding contributes to straighter teeth.

Items 42 to 46 are based on the following diagram.

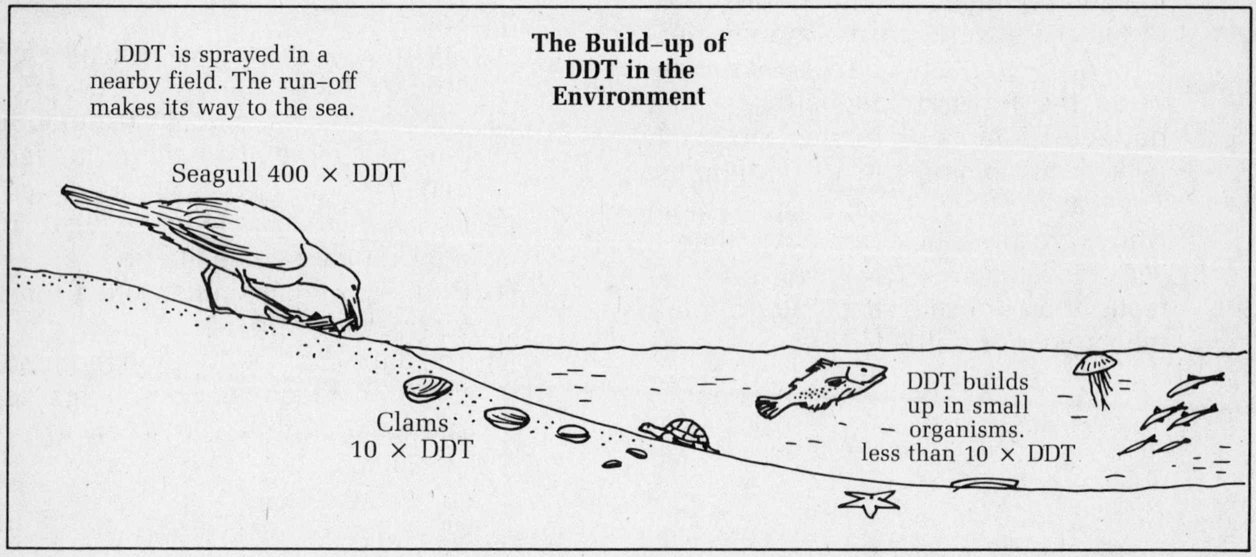

42. What relationship is demonstrated between what an animal eats and the amount of DDT in its body?

(1) Animals higher up on the food chain have higher amounts of DDT in their bodies
(2) Animals higher up on the food chain have lower amounts of DDT in their bodies.
(3) Animals that eat seafood have higher amounts of DDT in their bodies.
(4) Seagulls have higher amounts of DDT in their bodies than do clams.
(5) Small organisms have lower amounts of DDT in their bodies than do fish.

43. Comparing body weights can help give an idea of what concentrations of DDT will be found in animals that eat DDT-contaminated food. Which of the following concentrations of DDT is most likely for a person who often eats DDT-contaminated clams for dinner?

(1) 2 × DDT
(2) 5 × DDT
(3) 10 × DDT
(4) 20 × DDT
(5) 400 × DDT

44. The use of DDT has been banned in the United States. Based on the information in the picture, what was the effect of this ban on the environment?

(1) DDT was eliminated immediately from living things.
(2) DDT is being eliminated gradually from living things.
(3) The insect population increased greatly.
(4) The insect population decreased greatly.
(5) The clam population increased greatly.

45. Which fact below does not support the view that the disadvantages of DDT outweigh the advantages?

(1) DDT weakens bird eggshells.
(2) DDT is stored in animal fat.
(3) DDT kills many insects.
(4) DDT builds up in the environment.
(5) DDT appears in mother's milk.

46. Which of the following statements is a conclusion that can be drawn from the information in the picture?

(1) DDT build-up in the environment kills seagulls.
(2) Sprayed DDT affects only clams, seagulls, and insects.
(3) DDT run-off is absorbed by the sea, where it becomes harmless.
(4) Seagulls have adapted to DDT because they continue to thrive.
(5) DDT use affects the entire surrounding ecosystem.

---

47. Warm air can hold more water vapor than cold air. When air of a certain temperature is holding all the water vapor it can, it is said to be saturated. The saturation temperature is known as the dew point, because water will precipitate out of air cooled below that point. This is why you can see your breath on a cold day—the cold air cannot hold all the water vapor in your breath. The water vapor precipitates into droplets of water that you can see.

A cool, cloudless morning follows a hot, humid summer night. When you step outside that morning, the grass is covered with

(1) dew.
(2) frost.
(3) ice.
(4) rain drops.
(5) snow.

Item 48 is based on the following sketch.

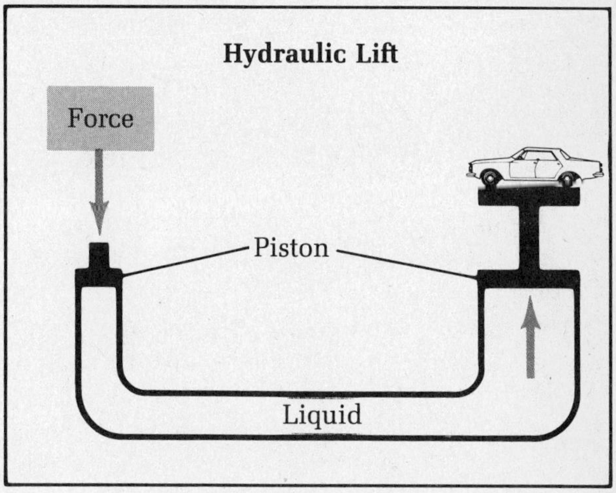

48. A small increase in the downward force on the smaller piston would cause

(1) the car to move downward.
(2) the car to move upward.
(3) no change in the car's position.
(4) the liquid to spill over the sides.
(5) the liquid to expand.

Items 49 to 53 refer to the following passage.

The sea lamprey is a fish that invaded the Great Lakes after the Erie Canal opened a passageway between the lakes and the Atlantic Ocean. The first sea lamprey in the Great Lakes was recorded in 1835, 10 years after the canal opened. However, it wasn't until the late 1800s that commercial fishers began to notice large numbers of sea lamprey scars on fish in Lake Ontario.

The adult sea lamprey is an eel-shaped fish with a round, open mouth. The lamprey uses its suckerlike mouth and teeth to attach to the side of fish, such as mackerel, salmon, and trout. The lamprey then feeds on the blood and muscles of its prey. The lamprey is aided by glands that secrete a substance that prevents its victim's blood from clotting. A lamprey can kill up to 40 pounds of fish in its two-year adult stage.

By 1946, the sea lampreys had populated all the Great Lakes. As sea animals, they had no natural enemies in the lakes, thus their population increased rapidly. Before the arrival of the lamprey, fishers in the upper lakes took in about 14 million pounds of trout per year. Thirty years later, the 1960 catch was only 0.5 million pounds. Factors besides the lamprey were partly responsible for this decrease—for example, commercial overfishing and pollution—but the lamprey played a large part.

Canada and the United States joined together in 1954 to control the sea lamprey. Early controls, such as mechanical and electrical barriers, did not work well. Next, a chemical was tried that kills young sea lampreys in their larval stage. Within 10 years, the lamprey population was reduced to about one eighth of its former size. The treatment is costly, however, and must be applied every four years to each of the more than 400 streams where sea lampreys spawn. Also, the chemical can kill other fish when concentrations become too high. Researchers now are looking for ways to control the population through the use of sterilized males or females.

49. In 1859 a farmer in Australia released on his land 24 rabbits from England. By the 1930s, the Australian rabbit population was estimated at 750 million. Based on the information regarding the sea lamprey's spread, what factor allowed the rabbit population to increase so quickly?

(1) The farmer fed the rabbits daily.
(2) The rabbits had no natural enemies in Australia.
(3) The rabbits ate as much grass and vegetables as they could.
(4) The rabbits were rapid breeders.
(5) The Australian continent has a small human population.

50. Which of the following is most likely to happen if a few sea lamprey larvae survive the pesticide used to kill them?

(1) The survivors will be less likely to kill other fish.
(2) The survivors will move to another stream.
(3) As the survivors reproduce, more lampreys will be immune to the chemical.
(4) The adults will avoid the poisoned areas when laying eggs.
(5) The survivors will die as they become adults.

51. Which of the following is not a disadvantage of using mechanical barriers to trap adult lampreys going upstream to spawn?

(1) The barriers are damaged easily by floating debris.
(2) The lampreys swim around them when the water is high.
(3) Lamprey larvae can swim through most barriers.
(4) The barriers trap fish other than lampreys.
(5) The barriers allow water and small fish to pass through.

52. The chemical used to kill sea lamprey larvae is a phenol, a family of chemicals that is toxic for a long time in the environment. Based on this information and that in the passage, which of the following is the most likely outcome of long-term use of this pesticide?

A. Other fish will begin to die as concentrations increase.
B. The Great Lakes will contain a measurable amount of the chemical.
C. The chemical's use will decrease because less will be required for the same effect.

(1) A only
(2) B only
(3) C only
(4) A and B only
(5) B and C only

53. Which of the following statements is evidence that sea lampreys are responsible for a large part of the decline in Great Lakes' fish catches?

(1) As lampreys increased, other fish decreased.
(2) The Great Lakes region contains much heavy industry.
(3) The Great Lakes are popular fishing areas.
(4) The lampreys have no natural enemies in the Great Lakes.
(5) The lampreys are voracious eaters.

---

54. Scientists, trainers, and doctors disagree on how much and whether a person should stretch before exercising. Those who are pro-stretching reason that a short muscle lacking in flexibility is more likely to be injured than a long, flexible one. Others, however, are becoming convinced that more people injure themselves while stretching than because of lack of flexibility.

Which of the following test results would lend support to those who believe that a lack of flexibility leads to more sports injuries?

(1) Athletes' muscles proved to be more flexible than those of nonathletes.
(2) Joggers who did not limber up before running had more injuries than those who did.
(3) Football players had more injuries than joggers.
(4) Dancers who limbered up first had fewer injuries than skiers who did not.
(5) Child athletes had fewer injuries than adult athletes.

55. Any process that causes part of the Earth's upper mantle to melt will create the kinds of lava that erupt quietly and form basalt, a dark-colored rock. This type of volcano is found where plates under the Earth's crust are separating or along major breaks in the Earth's surface. Along boundaries where plates are coming together, volcanoes tend to erupt violently. Their lava forms light-colored rocks.

Which of the following statements best explains why there are two different types of volcanic eruptions?

(1) Differences in the two types of lava create pressure differences.
(2) Differences in pressure are created depending on whether the Earth's surface is coming together or separating.
(3) The differences are due to the volcanoes' location on Earth.
(4) The differences are due to different kinds of volcanic gas.
(5) Differences in the lavas' colors cause a heat difference that influences the eruption's force.

Items 56 to 59 refer to the following information.

Chemical properties are determined by how a substance acts under certain conditions. Below are five chemical properties.

A. combustibility: how easily the substance burns

B. ionization: the degree to which the substance breaks up into electrically charged particles in solution

C. relative activity: how easily the chemical reacts compared to other members of its chemical family

D. stability: the degree of resistance of a substance to breaking down into other substances

E. test reactivity: how the substance reacts with certain test substances, such as water or litmus paper

Each of the above items might be used more than once in the following set of items.

56. Salt separates into positively and negatively charged particles when dissolved in water. Salt shows a high degree of which chemical property?

(1) combustibility
(2) ionization
(3) relative activity
(4) stability
(5) test reactivity

57. Children's pajamas are treated with a substance that reduces the fabric's ability to catch fire. The treatment results in the reduction of which chemical property?

(1) combustibility
(2) ionization
(3) relative activity
(4) stability
(5) test reactivity

58. To separate water into its hydrogen and oxygen parts, energy must be added. Water has a fairly high degree of which chemical property?

(1) combustibility
(2) ionization
(3) relative activity
(4) stability
(5) test reactivity

59. Solid sodium hydroxide is a drain cleaner that releases a large amount of heat and also fizzes vigorously when it comes into contact with water. In water, sodium hydroxide shows a high degree of which chemical property?

(1) combustibility
(2) ionization
(3) relative activity
(4) stability
(5) test reactivity

Item 60 refers to the following graph.

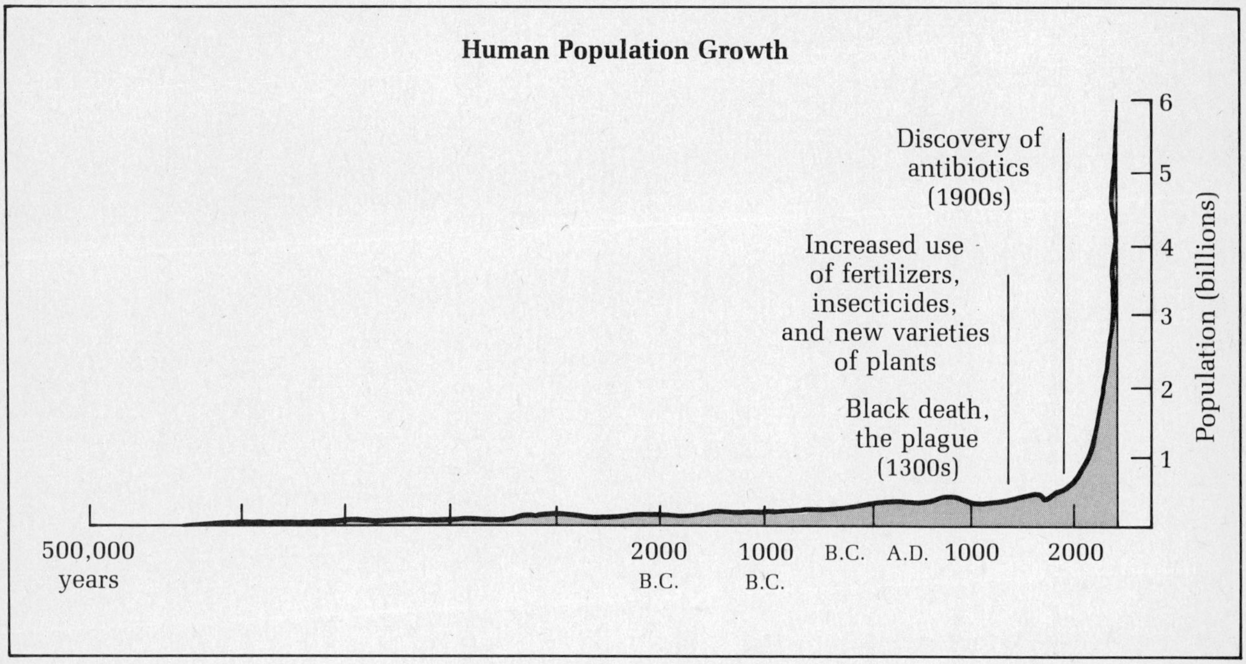

60. Which of the following is a hypothesis that could be used to explain the fact that the human population has increased so dramatically in the past few hundred years?

(1) The incidence of fatal diseases, such as the plague, has decreased.
(2) The use of fertilizer has increased the available food supply.
(3) Humans have eliminated or reduced those factors that slowed growth in the past.
(4) The birth rate has increased greatly, and the death rate has decreased greatly.
(5) The development of new technology has increased at a rapid speed.

---

61. The United States Department of Agriculture (USDA) grades beef according to the following system. Prime beef is tender, juicy, and flavorful because it has abundant marbling, or flecks of fat; Prime meats are more expensive than other grades. Choice beef has slightly less marbling than Prime and is slightly less expensive. Good beef is not as juicy or flavorful as the other two because it contains less fat; it is the least expensive of the three grades.

On what basis would a person trying to lose weight choose to eat only "Good" grade beef?

(1) Good beef has less flavor; the dieter would want to eat less of it.
(2) Good beef costs less, so the dieter can buy more of it.
(3) Good beef has less fat and therefore has fewer calories.
(4) Good beef is better tasting than lower grades of beef.
(5) Good beef is better for low-calorie cooking methods, such as broiling or grilling.

Items 62 to 64 refer to the following diagrams.

A.

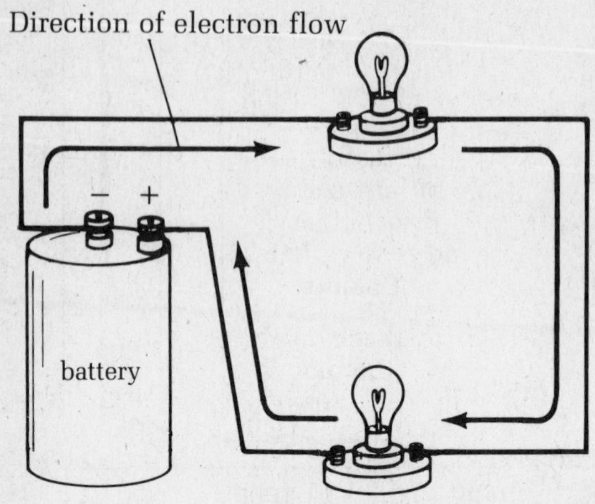

Series Circuit

B.

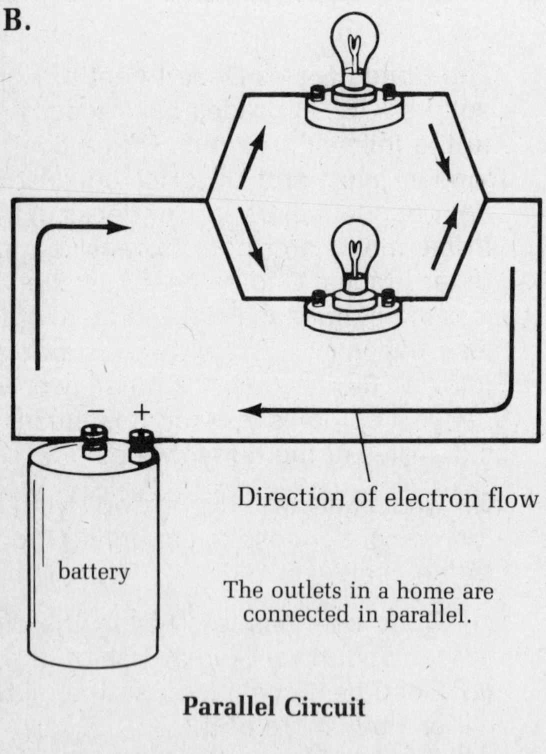

Parallel Circuit

62. Based on the diagrams, which of the following statements is always true of both series and parallel circuits?

(1) The electrons always flow from negative to positive terminals.
(2) The electrons always flow from positive to negative terminals.
(3) The electrons follow one path only.
(4) A break at any point in the circuit will disconnect the entire circuit.
(5) If one light bulb burns out, all the others will burn out too.

63. The outlets in your home are connected in parallel. If they were connected in series, what would happen when you turned off one lamp?

(1) The lights nearby would go dark also.
(2) All the appliances on the circuit would go off also.
(3) The appliances that require the most electricity would go off also.
(4) The appliances that require the least electricity would go off also.
(5) Only the light you turned off would go dark.

64. To protect a circuit from overloading with electricity, fuses or circuit breakers break the flow of electricity when it reaches a dangerous level. At which of the following locations in a home would circuit breakers or fuses be most effective?

(1) near the largest number of appliances
(2) near appliances requiring the most electricity to run
(3) in the middle of the house
(4) near the power source
(5) far away from the power source

65. Skin cancer is the most common form of cancer in both women and men. Several factors influence the risk of skin cancer. Overexposure to the sun is the most common cause. Heavy exposure to radiation, such as x rays, also can cause skin cancer. A third factor is skin pigment—people with light skin who burn easily are most at risk for skin cancer. Based on this information, which of the following individuals is most at risk for skin cancer?

(1) a dark-skinned man who is a construction worker
(2) a dark-skinned woman who has dental x rays taken every year
(3) a light-skinned dental assistant who takes x-ray pictures almost every day
(4) a light-skinned man who jogs outdoors for exercise after work
(5) a light-skinned man who is a professional fisherman

66. A recent experiment sought to explain why people who read a great deal often become nearsighted. The researchers discovered that the black and white pages deprive eye cells of the stimulation of different sizes, various shapes, and colors found in the real world. This deprivation might encourage the change in the eye shape that leads to nearsightedness.

Based on this research, how might publishers design pages differently to reduce the incidence of nearsightedness? They should design pages with

(1) colored letters of the same size.
(2) larger black letters.
(3) various-sized colored forms along with the usual black letters.
(4) various-sized black letters.
(5) smaller black letters with a few words in color.

Check your answers on pages 134–136.

SIMULATED PRACTICE TEST

Performance Analysis Chart

Directions: Circle the number of each item that you have correct on the Simulated Practice Test. Count how many items you have correct in each row; count how many items you have correct in each column. Write the amount correct per row and column as the numerator in the fraction in the appropriate "Total Correct" box. (The denominators represent the total number of items in the row or column.) Write the grand total correct over the denominator **66** at the lower right corner of the chart. (For example, if you have 60 items correct, write *60* so that the fraction reads *60*/**66.**) Item numbers in boldface represent items based on graphic material.

Item Type	*Biology (Chapter 2)*	*Earth Science (Chapter 3)*	*Chemistry (Chapter 4)*	*Physics (Chapter 5)*	TOTAL CORRECT
Comprehension	9, 14, 22, 23, **42,** 65	2, **10,** 35, 36		16, **48**	**/12**
Application	3, 4, 5, 6, 7, **15,** 21, **43,** 49, 66	**11,** 47	8, 20, **29,** 56, 57, 58, 59	1, 17, 18, **63, 64**	**/24**
Analysis	24, 25, 31, **40,** 41, **44,** 50, 51, 52, **60**	**12, 13,** 37, 38, 55	**30,** 34	28	**/18**
Evaluation	26, 32, **45, 46,** 53, 54, 61	**27,** 39	33	19, **62**	**/12**
TOTAL CORRECT	**/33**	**/13**	**/10**	**/10**	**/66**

The chapters named in parentheses indicate where in the Science instruction of the Cambridge Comprehensive Program (Unit V) and the Cambridge Program for the Science Test (Unit II) you can find specific instruction about each area of Science.

On the chart, the item types are classified as Comprehension, Application, Analysis, and Evaluation. For reinforcement of the instruction for these problem types, refer to the appropriate sections of Unit III of the Comprehensive Book and Unit I of the Program For the Science Test.

Answers and Explanations

Biology

1. **(3)** *Comprehension/Biology.* A clone is by definition an identical organism descended from a single parent or ancestor.

2. **(3)** *Application/Biology.* Although a clone is an exact copy of its ancestor and thus looks like it, the fact that an organism looks like one of its parents does not make it a clone, because it has two parents.

3. **(4)** *Evaluation/Biology.* The other choices are all examples of cloning that occurs in nature without human intervention.

4. **(4)** *Comprehension/Biology.* Heat, gravity, and the chemical (water) are the same for both positions, eliminating these as choices. The position of the smooth surface changes, but first the shrimp is swimming with its back toward the smooth surface, then with its front toward the smooth surface. In both positions, the shrimp is swimming with its front toward the light, making the light the most likely stimulus.

5. **(5)** *Comprehension/Biology.* If gibberellic acid decreases the amount of grain produced by a plant, the plants with low levels of gibberellic acid will tend to produce more grain than will plants with high levels of the acid. The increased amount of grain per plant is an advantage to a farmer. Choice (1) is most likely true of plants that produce high levels of the acid. There is no basis for selecting choices (2), (3), or (4).

6. **(4)** *Analysis/Biology.* In order to reach the conclusion, one must assume that A human cancer cells will be blocked by the antitumor compounds in the same way the rodent cancer cells were. One also must assume that C the compounds work when eaten and digested in a normal way, rather than isolated from the beans, rice, and potatoes. It is not necessary to assume B, because the question is concerned only with human cancers, not all cancers.

7. **(5)** *Comprehension/Biology.* The shortest plant on the graph is the one grown outside in full sunlight. Choices (1), (2), and (3) are not addressed by the graph.

8. **(2)** *Comprehension/Biology.* The tallest plant, Plant A, was grown indoors. A is not addressed by the graph. C is false, according to the graph.

9. **(5)** *Analysis/Biology.* Both B and C are logical, according to the graph. A is illogical, because a plant in the shade receives less sunlight and thus less energy.

10. **(2)** *Evaluation/Biology.* Soil, species, light intensity and water can all affect plant growth. The color of the plant flower would have no effect on plant growth.

11. **(2)** *Evaluation/Biology.* The host (human) in the example is harmed but not killed. The parasite (trichina worm) benefits by being fed and sheltered by the host.

12. **(4)** *Analysis/Biology.* The paragraph states that flu-like symptoms are common with trichinosis, but the facial swelling is the real tipoff. If the larvae invade the muscles of the face and neck, the resulting swelling would signal something other than flu. Choice (1) is incorrect because the symptoms would not appear for at least 16 days, after the worms mature and reproduce. Choice (3) is incorrect because trichinosis is not contagious. Choice (5) might enter the doctor's mind but would not be the determining factor in his diagnosis.

13. **(3)** *Application/Biology.* The plant is most likely to flower in the summer, when the days are long and the nights short.

14. **(1)** *Analysis/Biology.* Photoperiodism is a response to change over the period of a year. The other choices are processes that occur at certain times of day throughout the growing season.

15. **(5)** *Application/Biology.* Tropical rain forests contain vines that grow on other plants, as well as sufficient moisture in the air for a plant to obtain water through airborne roots.

16. **(5)** *Comprehensive/Biology.* Tropical rain forests receive 200 to 400 centimeters of rain a year, an indication of frequent rain.

17. **(2)** *Application/Biology.* The dominant vegetation in the grassland biome is grasses and wild flowers. Of the choices, only buffalo and zebras feed on grasses. The zebra is eliminated as a choice because it is not a North American animal; it is a native of Africa.

18. **(2)** *Analysis/Biology.* According to the table, rainfall in the two biomes is about the same, eliminating choice (1). Choices (3), (4), and (5) are not covered by the table. However, according to the table, taiga is located in latitudes farther from the equator than grasslands. From this it can be deduced that the grasslands tend to have warmer temperatures than the taiga, which could account for the difference in vegetation.

19. **(2)** *Analysis/Biology.* The desert biomes are located fairly near to the equator and receive little rain. Those animals that get water from plants would tend to survive.

20. **(1)** *Analysis/Biology.* Green plants are the main link in the food chain. Where there are more green plants, there are more animals. Choices (2) and (3) are untrue as shown by the chart. Choices (4) and (5) are irrelevant.

21. **(4)** *Evaluation/Biology.* The paragraph states that as one part of the ecosystem (plants and bacteria) increased, another part (the fish) showed a rapid decrease. With its balance disturbed, the entire ecosystem's existence was threatened.

22. **(3)** *Application/Biology.* People eat vegetables, fruits, grains, and meat, a diet that consists of both plants and animals.

23. **(5)** *Application/Biology.* Fallen leaves, logs, and animal droppings are decaying matter from plants and animals, classifying the mushroom as a saprophyte.

24. **(4)** *Application/Biology.* The maple tree makes its own nutrients, classifying it as a producer.

25. **(2)** *Application/Biology.* Hay, oats, grass, carrots, and sugar are plant foods, classifying the horse as an herbivore.

26. **(1)** *Application/Biology.* The lion feeds on other animals, classifying it as a carnivore.

27. **(1)** *Application/Biology.* Insects are animals, classifying the spider as a carnivore.

28. **(3)** *Evaluation/Biology.* The graph shows that eggshells became much thinner following the first widespread use of DDT. Thus the best hypothesis should include this fact, making choice (3) the correct one. The graph contains no evidence for choices (1), (4), and (5). Choice (2) is false, according to the graph.

29. **(4)** *Analysis/Biology.* Statement A is an advantage, because it simplifies the farmer's work. Statement C could work for or against the farmer, depending on the selling price of the crop when it is marketed. Statements B and D are disadvantages that the farming of several crops at once would alleviate.

30. **(5)** *Application/Biology.* Each of the other choices can be controlled to some extent; however, the dilation and constriction of our pupils in response to light are not under our conscious control.

31. **(1)** *Application/Biology.* The other choices would reduce the amount of fat consumed, which studies show helps reduce the risk of heart disease. Eating less cereal, however, probably would mean consuming less grain, which goes against a recommended dietary change.

32. **(4)** *Evaluation/Biology.* The passage states that eating less cholesterol reduces cholesterol in the blood, thereby reducing the risk of heart disease; hence the AHA recommendation.

33. **(1)** *Analysis/Biology.* Only choice (1) can be proven by scientific experimentation.

34. **(2)** *Evaluation/Biology.* Each time the hare population declines, the lynx population declines soon after. Although this occurs about every ten years, it is not logical to assume this is true of all prey-predator relationships, eliminating choice (3). Choice (1) is false, according to the graph. Choice (4) also is false, because both species show a cyclical decline followed by an increase, indicating neither species is harmed by overcrowding, although individual animals might die as a result. Choice (5) is not addressed by the graph.

35. **(3)** *Application/Biology.* The noise level in most factories is extremely high. The other choices describe situations that do not produce extreme noise, even choice (1), because passengers in an airplane are not subjected to the noise levels that the ground crew on the runway experiences.

36. **(2)** *Application/Biology.* Only choice (2) is an automatic, inherited response.

37. **(5)** *Comprehension/Biology.* An inherited automatic behavior is not learned.

38. **(2)** *Analysis/Biology.* Effective, and therefore, valuable behavior to an animal is one that increases its chances for survival. The ability to vary behavior might increase any of the factors listed at one point or another, but overall it must increase the animal's chances for survival to be of value.

39. **(2)** *Evaluation/Biology.* A newly hatched duckling that follows anything is exhibiting inherited behavior; once the duckling has imprinted on the object, the behavior cannot be changed.

40. **(1)** *Analysis/Biology.* According to the passage, humans must learn nearly every behavior. Learning generally takes place in the brain, so it is likely that this body part has the greatest influence on human behavior.

41. **(3)** *Comprehension/Biology.* The figure shows that the earthworm and woodlouse feed on plant remains, and the maggot feeds on animal remains.

42. **(4)** *Analysis/Biology.* Food energy is lost at each step up a food chain. Thus an animal large enough to kill an animal, such as a fox, found at the end of such a chain needs a lot of energy. It is more energy-efficient for an animal to eat lower down on the food chain.

43. **(2)** *Evaluation/Biology.* Choices (1), (3), and (4) are false, according to the figure. Choice (5) is true of this figure, but other food chains might have large animals near the beginning. A cow, for instance, feeds on grass and then is fed on by other, often smaller, organisms, such as human beings.

44. **(5)** *Application/Biology.* In the figure, the arrow points toward the predator or consumer. In this particular food chain the only arrangement that makes sense is one in which the chicken eats the grain and in turn is eaten by the person, as shown by the arrows.

45. **(3)** *Application/Biology.* According to the information provided, each step up the food chain results in a loss of food energy. Thus the longest chain would lose the most food energy.

46. **(1)** *Comprehension/Biology.* A chemical that lures male moths to female moths is most likely a sex attractant.

47. **(2)** *Application/Biology.* Only choice (2) would not necessitate an adaptation. The remaining choices involve environmental changes that would threaten the population and necessitate an adaptation.

48. **(1)** *Analysis/Biology.* Because conditions were the same for both sets of seeds, and more seeds germinated with the treatment than without it, it can be concluded that the treatment had a positive effect on seed germination. Choices (4) and (5) would have to be tested independently.

49. **(5)** *Application/Biology.* The chloroplasts are making two simpler substances (carbon dioxide and water) into glucose.

50. **(2)** *Application/Biology.* The lysosomes are digesting food substances into a usable form for the cell.

51. **(1)** *Application/Biology.* The cell membrane is involved in the passage of materials into the cell.

52. **(3)** *Application/Biology.* The genes are involved in the production of new cells.

53. **(5)** *Application/Biology.* The endoplasmic reticulum is involved in making new substances.

54. **(1)** *Analysis/Biology.* Because vitamin A is found in carrots, feeding infection-prone children extra carrots and monitoring the results would test the theory that extra vitamin A reduces infection.

55. **(5)** *Analysis/Biology.* The experiment would be worthless if medical care varied for the children, because the care they received could affect the outcome. Choice (1) is incorrect because adding <u>extra</u> vitamin A to the diet assumes the children already are receiving <u>some</u> vitamin A. Choices (2) and (3) are incorrect because the number of infections a child contracts doesn't depend on age or sex. Choice (4) is incorrect because some of the children will become ill during the course of the experiment; the experiment is testing whether the vitamin A will reduce the number of infections, not eliminate them altogether.

56. **(4)** *Comprehension/Biology.* According to the graph, the rate of growth is greatest between 30° and 40° Centigrade.

57. **(2)** *Comprehension/Biology.* A renewable resource is one—such as a guayule plant or rubber tree—that can be replanted and replaced when used. An example of a nonrenewable resource is petroleum, which takes millions of years to form and thus cannot be replaced easily.

58. **(3)** *Evaluation/Biology.* A race car driver would be most concerned with having a tire that lasted as long as possible and would resist blowouts, because dependable equipment is important to his or her livelihood.

59. **(5)** *Analysis/Biology.* Such a group would be concerned with the effects on the environment and thus would support natural rubber production mainly because it does not cause pollution.

60. **(1)** *Analysis/Biology.* A farmer would be most interested in finding a crop that grows well in his or her particular area.

61. **(5)** *Analysis/Biology.* For simple molecules to have made complex molecules, the simple molecules had to have come first. Choice (1) is irrelevant to the theory. Choices (2) and (3) are stated in the theory. Choice (4) might or might not be true, but it is not a necessary part of the theory.

62. **(4)** *Application/Biology.* By studying the figures, you can see that a plant's roots grow down, and a plant's stem grows up, regardless of the tilt of the pot the plant is in. Applying this to the situation described, the flower's stem and roots will most resemble those in Figure B.

63. **(2)** *Analysis/Biology.* Choices (1), (3), and (5) are statements of fact from the study, not conclusions. Choice (4) is an assumption that might or might not be true. The only conclusion listed is choice (2).

64. **(3)** *Analysis/Biology.* The roots probably were damaged when the tree was dug up and transplanted. Because the tree takes in water through the roots, it no longer received enough water, causing its leaves to shrivel and the tree to die.

65. **(4)** *Comprehension/Biology.* All of the other choices describe species that might live in or on a grass lawn. A fish-eating species, however, would not find food in this habitat.

66. **(1)** *Analysis/Biology.* The human species has proved itself able to compete with every species on Earth, eliminating choices (2) and (3). We live on mountains and in deserts, eliminating choices (4) and (5). Only lack of more space in which to expand is a limiting factor (out of the choices given).

67. **(5)** *Evaluation/Biology.* According to the passage, at early stages in its development, a species is found in just a few areas. Thus the fact that the earliest human remains are found in Africa is evidence that the species evolved on that continent first.

68. **(2)** *Evaluation/Biology.* The passage states that a species' evolutionary life lasts an average of one million years. Therefore, it follows logically that the species will not live much longer in its current form. There is no logical foundation for any of the other choices.

69. **(1)** *Analysis/Biology.* The time period and fossil record suggest that Equus evolved from Eohippus. Choices (2), (3), and (4) are facts supporting the hypothesis. Choice (5) is incorrect because Equus lives in grasslands and Eohippus lived in the jungle and thus fed on different foods and occupied a different niche.

70. **(4)** *Application/Biology.* A bee is an insect and thus is likely to be a member of a social group. The other choices are neither insects nor vertebrates.

71. **(1)** *Comprehension/Biology.* The paragraph implies that animals can move around on their own and must obtain and digest food from the outside environment, traits most protozoa share. Animals, however, are multicelled organisms, whereas protozoa are made up of one cell.

72. **(3)** *Application/Biology.* Vitamin C is the only vitamin in the table that aids iron absorption. Of the list of fruits, lemons are the only citrus fruit and thus the fruit that contains a significant amount of vitamin C and that will aid in iron absorption.

73. **(4)** *Analysis/Biology.* According to the table, vitamin D aids in the formation of healthy bones, and a good source of vitamin D is sunshine. Putting these two pieces of information together with that in the item leads to the conclusion that rickets is caused by a lack of vitamin D.

74. **(4)** *Evaluation/Biology.* The symptoms of scurvy listed in choices (1), (2), (3), and (5) are what you might expect from a lack of vitamin C, given its function in the table. However, another deficiency or disease could cause these symptoms as well. Only choice (4) provides a direct link between vitamin C and scurvy by adding fruits high in vitamin C to the diet.

75. **(4)** *Evaluation/Biology.* Priestly proved van Helmont wrong by showing that the tree took in carbon dioxide as well as water. Carbon dioxide comes from the air, and thus the tree used water and air to grow. The other choices do not follow logically from the passage.

76. **(2)** *Comprehension/Biology.* The passage states that the more people there are, the more ecosystems are affected. It gives specific examples of ecosystem destruction as covered in the other choices, but the ultimate cause is the rapid growth of the human population.

77. **(5)** *Analysis/Biology.* The passage states that highly developed countries like the U.S. cause air pollution and that the U.S. burns a great deal of fossil fuels, which give off poisonous gases. Putting these two facts together leads to the conclusion in choice (5).

78. **(3)** *Analysis/Biology.* Measure A would help, because fossil fuels give off poisonous gases. Measure B would help reduce pollution from factories. Measure C would help reduce both solid waste pollution and the amount of raw materials taken from the environment to make these products. Only measure D would not help, because creating farms destroys the natural ecosystems in these areas.

79. **(1)** *Evaluation/Biology.* Penicillin is a beneficial drug that was derived from a seemingly insignificant living organism and thus supports the argument. The other choices list products or organisms of value to humans but are not instances in which we learned something of value.

80. **(4)** *Application/Biology.* Alternatives A and B are both situations that normally cause mental stress. Neither C nor D would cause undue stress and would not affect the white blood cell count.

Earth Science

1. **(2)** *Comprehension/Earth Science.* An upwelling is initiated when wind stirs the surface océan water and blows it out to sea. Choice (1) indicates factors that are involved in upwellings but do not cause them. Choice (3) is incorrect because the mixing of cold and warm water is the result, not the cause, of an upwelling. Choices (4) and (5) indicate factors that are affected by upwellings but do not cause them.

2. **(4)** *Analysis/Earth Science.* Because upwellings moderate the temperature of warm surface water, a failure of upwellings to occur could have a destructive effect on sea life. Choices (1) and (5) are illogical because upwellings provide needed nutrients and optimum temperatures for fish. Choice (2) cannot be concluded because there has been no discussion of differences among sea creatures. Choice (3) is incorrect because upwellings bring colder water to warm surface water.

3. **(3)** *Comprehension/Earth Science.* In the passage, the term "greenhouse effect" is used to describe the warming of the Earth by gases in the atmosphere. Choices (1) and (5) are relevant to an actual greenhouse but not to the use of the term "greenhouse effect" in the passage. Choices (2) and (4) are irrelevant.

4. **(4)** *Evaluation/Earth Science.* Because carbon dioxide is the gas that traps most of the Earth's infrared rays, it is logical to assume that the more carbon dioxide there is in a planet's atmosphere, the more heat will be absorbed. Choices (1) and (2) are incorrect because the greenhouse effect will be operative as long as some carbon dioxide is present. Choice (3) contradicts the correct answer. Choice (5) is incorrect because ultraviolet radiation can penetrate carbon dioxide.

5. **(1)** *Analysis/Earth Science.* Because the burning of coal produces carbon dioxide, choice (1) is a statement of fact. Choice (2) is an opinion because it is the belief of many scientists—but not a proven fact—that the Earth's temperature will increase in the future. Choices (3), (4), and (5) are clearly statements that cannot be proven or measured.

6. **(5)** *Application/Earth Science.* Because ultraviolet radiation can pass through carbon dioxide in the Earth's atmosphere and infrared radiation cannot, choice (5) is a logical experimental procedure. Choice (1) is irrelevant. Choices (2), (3), and (4) are incorrect because both types of radiation can pass through glass and both provide heat.

7. **(3)** *Analysis/Earth Science.* Because warm air can hold more moisture than cool air, warmer midday air will reabsorb the fog. Choices (1), (2), and (4) are not supported by the information provided. Choice (5) is an illogical and meaningless mixture of information.

8. **(3)** *Comprehension/Earth Science.* The article clearly states that the gold deposit in Connecticut has a concentration that is much greater than that normally found in North America and greater even than that found in the world's leading gold mines. Nothing in the article indicates that it is especially unusual for gold to be found in a state forest (1), in an abandoned cobalt mine (2), near a city (4), or to be discovered accidentally (5).

9. **(5)** *Evaluation/Earth Science.* An environmental scientist is concerned about the preservation of natural resources, including land, water, plants, and animals. Choices (1) and (2) would not be of particular interest to a scientist. Choices (3) and (4) might be of some interest to an environmental scientist but would not be the *most* important factors.

10. **(5)** *Analysis/Earth Science.* Although the concentration of gold per ton of rock is clearly stated, the actual amount of gold—how much gold the deposit actually would yield—is not discussed. One rock with a lot of gold would have a high concentration of gold but hardly enough gold to supply the world. Choices (1), (2), and (4) are incorrect because this information is provided by the article. Choice (3) might be important in considering whether the state wants to mine the gold but not particularly relevant to whether the state can become a major world supplier.

11. **(1)** *Application/Earth Science.* Because gold concentration is given in ounces of gold per ton of rock, it is logical to assume that silver concentration would be expressed in the same way. The article clearly states that this concentration is determined by examining rock samples. Choice (2) might work, but it is not the best answer, based on the article. Choice (3) is impractical. Choices (4) and (5) are irrelevant to the definition of concentration.

12. **(2)** *Application/Earth Science.* The burning of the candle will use up the available oxygen in the air, causing the volume of the air to decrease according to the amount of oxygen present. Choices (1), (3), and (4) describe the correct experimental procedure, but the change in volume of the candle is irrelevant, and changes in mass are not what the scientist wishes to determine. Choice (5) is incorrect in all aspects.

13. **(4)** *Evaluation/Earth Science.* Because S waves cannot travel through liquids, following their course through this region would determine whether the rock is liquid or solid. Choices (1) and (2) are not relevant to the information provided. Choice (3) is incorrect because P waves travel through solids and liquids. Choice (5) is incorrect because L waves travel only along the Earth's surface, not in the inner Earth.

14. **(1)** *Analysis/Earth Science.* Choice (1) is correct because the apparent brightness of two stars will depend on their distances from an observer on Earth, whereas the apparent sizes of two objects in a photograph will depend on their distances from the camera. Choice (3) is similar to the correct answer, but it does not include a comparison of two objects. Choices (2) and (4) are tempting answers because they deal with light, but they are not relevant to the information given. Choice (5) is irrelevant also.

15. **(4)** *Comprehension/Earth Science.* A mineral can scratch any mineral below itself on the scale and can be scratched by any mineral above itself on the scale. Talc (1) and gypsum (2) could not scratch calcite, and fluorite (3) could not scratch apatite. Diamond (4) could not be scratched by any other mineral.

16. **(3)** *Application/Earth Science.* A mineral cannot scratch a mineral equal to itself, and thus a diamond could not be scratched by another diamond. Choice (1) is incorrect because the stone could have a hardness of 9 and still not be a diamond. Choice (2) is irrelevant. Choice (4) is incorrect because it would prove that the stone is a diamond. Choice (5) is incorrect because no mineral can scratch a diamond.

17. **(2)** *Evaluation/Earth Science.* Because it takes 6 billion years for light to travel 6 billion light years, the light from Star X would have had to begin its journey to Earth before the Earth was formed 4.6 billion years ago. The other choices would not indicate relative ages of stars to the Earth.

18. **(4)** *Comprehension/Earth Science.* The chart shows that the comfort index is based on humidity and temperature. Choice (1) is incorrect because although the comfort index indicates the percentage of people who feel comfortable, it is not <u>based</u> on this statistic. Choices (2), (3), and (5) are either incomplete or inaccurate.

19. **(2)** *Analysis/Earth Science.* Because the comfort index is based on both temperature and humidity, it is logical to conclude that the difference in comfort in these two cities is based on differences in humidity. The other choices might be true, but they are not addressed by the chart.

20. **(3)** *Comprehension/Earth Science.* The main focus of the article is concern that ocean pollution might be causing the dolphin deaths. Choices (1) and (4) might be true, but they are not relevant to the article. Choice (5) is implied by the article, but it is not the main focus of the article. There is no basis for selecting choice (2).

21. **(5)** *Analysis/Earth Science.* Because the lesions appear to be caused by acid, a possible explanation for the lesions is that acidic substances have entered ocean water. Choices (1), (2), and (4) represent facts that do not explain the cause of the lesions. Choice (3) represents an opinion, not a hypothesis.

22. **(1)** *Evaluation/Earth Science.* Choice (1) is correct because the article emphasizes concern about the link between the dolphin deaths and the safety of ocean water for human beings. Choice (2) is incorrect because no mention is made of the type of infection contracted by the swimmers. Choice (3) is not supported by the passage. Choice (4) is incorrect because the link between acid waste and the dolphin deaths has not been established. Choice (5) is incorrect because there is no evidence to support it.

23. **(4)** *Application/Earth Science.* Choice (4) is the most logical first step because examining the bodies of the dead fish might shed light on the cause of the deaths. Choices (1) and (5) would not yield scientific data; choices (2) and (3) make sense only after the autopsies have given reason to believe that the deaths are linked to water pollution.

24. **(5)** *Analysis/Earth Science.* By the time horses appeared, the dinosaur had become extinct. All other choices represent aspects of life on Earth that could have existed at the same time.

25. **(3)** *Analysis/Earth Science.* The diagram shows that because the Earth is between the moon and the sun, the moon is in darkness. Choice (1) is incorrect because the moon could move out of the shadow of the Earth and still be just as far from the sun. Choice (2) is not shown by the diagram. Choice (4) describes an eclipse of the sun, not an eclipse of the moon. Choice (5) is a true statement that is irrelevant to the question.

26. **(5)** *Comprehension/Earth Science.* The two graphs show that both factors increase, but temperature increases most sharply just below the Earth's surface, whereas pressure increases least sharply between 0 and 2,000 km and more sharply between 2,000 and 6,000 km.

27. **(4)** *Evaluation/Earth Science.* Because the variable in this experiment is the angle of the light source, choice (4) is the best answer. Choices (1) and (2) are incorrect because water is present in equal amounts in both setups. Choice (3) is incorrect because the distance of the light source from the water is essentially the same in both setups. Choice (5) is irrelevant.

28. **(1)** *Analysis/Earth Science.* A higher Ep value would make the value of the P/Ep fraction smaller. Choice (2) is incorrect because it is the opposite of the correct answer. The other choices might be true, but they do not relate to the information provided.

29. **(2)** *Comprehension/Earth Science.* Because Oakee Pond is on the 200-ft. contour line and Galley's Hill has an elevation of 800 ft., the difference between the two points is 600 ft.

30. **(1)** *Comprehension/Earth Science.* Because a contour line exists between the 20-ft. lines numbered on the map, the interval between contour lines must be 10 ft.

31. **(4)** *Application/Earth Science.* The elevation from the shoreline to the Cabana Dunes changes gradually from 0 to 70 feet. The other choices describe the land surface incorrectly.

32. **(4)** *Analysis/Earth Science.* Because each contour line represents a change in elevation, contour lines close together indicate elevation that is changing very rapidly. Choice (1) is irrelevant. Choice (2) is incorrect because it contradicts the correct answer. Choice (3) is incorrect because a topographic map does not show population. Choice (5) is incorrect because it misinterprets the definition of a contour interval.

Chemistry

1. **(3)** *Comprehension/Chemistry.* It is clear from the values given in the chart that the boiling point of water increases as atmospheric pressure increases. Choices (1), (2), and (5) contradict the information presented. Choice (4) is meaningless.

2. **(2)** *Analysis/Chemistry.* At lower atmospheric pressure, water boils at a lower temperature. Thus the egg would be cooking at a lower temperature on a mountaintop and would be receiving less heat per unit of time. Choices (1) and (3) are incorrect because water actually would come to a boil more rapidly when the boiling point is at a lower temperature. Choices (4) and (5) are meaningless.

3. **(1)** *Comprehension/Chemistry.* When TNT explodes, new substances are produced, indicating that a chemical reaction has taken place. Melting point and freezing point—choices (2), (3), and (5)—are physical properties. Density—choice (4)—is also a physical property.

4. **(2)** *Application/Chemistry.* Because properties can be used to recognize or identify a substance, knowledge of the properties of gold—particularly its density—would be very useful in determining the authenticity of the ring. In choices (1), (3), (4), and (5), it would not be helpful to know the properties of the things in question.

5. **(1)** *Comprehension/Chemistry.* Statement A involves indirect evidence because the child must

learn about the object in the box without seeing it or touching it. Statement B does not involve indirect evidence because the man can observe the tire and watch air bubbles escape from the leak. Statement C does not involve indirect evidence because the woman can measure both the cloth and the sofa.

6. **(5)** *Evaluation/Chemistry.* The mixing of milk with vinegar to produce a curd parallels the statement that when milk reacts with an acid it produces a coagulated protein. The statement that vinegar is a coagulated protein is false because the coagulated protein is the product of the reaction, not the substance reacting with milk.

7. **(3)** *Comprehension/Chemistry.* Because the subscript following each symbol in a chemical formula represents the number of atoms, there are twice as many hydrogen atoms present as oxygen atoms.

8. **(4)** *Analysis/Chemistry.* Choice (4) is correct because silver tarnishes when brought into contact with a sulfur-containing substance. Choices (1) and (3) are not supported by the information given. Choice (2) is incorrect because it is not specific enough, and it jumps to the conclusion that egg yolks and rubber are the same in composition, which they are not. Choice (5) is incorrect because it offers no explanation for the presence of the sulfur.

9. **(3)** *Application/Chemistry.* Two electrons would fill level 1, then 8 electrons would fill level 2, then 1 electron would be left to occupy level 3. Level 3 would be the outermost energy level.

10. **(3)** *Application/Chemistry.* Two electrons would occupy level 1, leaving 5 electrons to occupy level 2. Level 2 would be the outermost energy level.

11. **(4)** *Analysis/Chemistry.* The passage states that the chemical properties of an element are closely related to the number of electrons in the outer shell. The other choices are all true statements, but they do not relate to the passage.

12. **(2)** *Analysis/Chemistry.* Because level 3 can hold only 18 electrons, this level could not receive an additional electron unless there were at least one empty space.

13. **(4)** *Comprehension/Chemistry.* The diagram shows that the surface of the liquid is viewed at eye level and the reading is taken at the lowest portion of the meniscus.

14. **(4)** *Comprehension/Chemistry.* If Tum-Aid reacts to remove acid, its active ingredient must be a base. Choices (1) and (3) are incorrect because these are products, not reactants, of the reaction to neutralize acid. Choice (2) is incorrect because the purpose of the medication is to reduce the amount of acid, not to increase it. Choice (5) is irrelevant.

15. **(2)** *Analysis/Chemistry.* By mixing Tum-Aid with the same type of acid found in the stomach, it is easy to determine whether or not a salt and water are produced. The remaining statements are subjective and cannot be proven or measured.

16. **(5)** *Analysis/Chemistry.* Because the neutralization of an acid produces salt, the reaction of Tum-Aid with stomach acid would produce a salt. Although the information provided does not specify that the salt produced is a sodium salt, this is still the most logical answer. Choice (1) is incorrect because the reaction between an acid and a base produces water—and this would not lead to dehydration. Choice (2) is a reason why you take Tum-Aid and should not be a side effect. There is no basis for choice (3), and choice (4) contradicts the idea that Tum-Acid reduces the amount of acid in the digestive system.

17. **(3)** *Evaluation/Chemistry.* Because Tum-Aid is designed to neutralize stomach acid, it is most likely that the medicine would fail to bring relief when the ailment is caused by some other factor. The other choices could be true, but they do not relate to the information provided. Choice (5) is rather unlikely, because product tampering usually results in the medication becoming harmful, not ineffective.

18. **(2)** *Comprehension/Chemistry.* Because the water is at the 250-cm^3 mark before the rock is immersed and at the 400-cm^3 mark after the rock is immersed, the volume of water displaced is 150 cm^3. Therefore, the volume of the rock must be 150 cm^3. If you answered choice (1), you confused the mass of the rock (50 grams) with the volume of the rock.

19. **(2)** *Analysis/Chemistry.* Magnesium oxide is produced when the oxygen and magnesium inside the flashbulb react. Because oxygen is a gas, the consumption of the oxygen would create a vacuum inside the bulb, resulting in reducing pressure and a slight collapse of the bulb. Choices (1) and (5) are plausible, but they do not relate directly to the information provided. Choice (3) is unlikely, because particles of a powder probably would exert very little force. Choice (4) also is unlikely, because an explosive reaction (if indeed this reaction is explosive) probably would cause a bulb to shatter, not collapse.

20. **(4)** *Application/Chemistry.* Making sawdust out of a log breaks a solid into very small pieces and increases its surface area. As a result, the chemical reaction of burning would be greatly speeded up. Choices (1) and (2) are incorrect because the amount of wood would be the same if the sawdust is made from the log. Choice (3) is incorrect because the match cannot be a catalyst—it is burned up, and the passage states that a catalyst emerges from a reaction unchanged. The passage gives no basis for choice (5).

21. **(1)** *Application/Chemistry.* Cutting the meat into several pieces would increase the surface area, thus causing the meat to cook faster. Choices (2), (3), and (5) are irrelevant. Choice (4) would cause the meat to cook more slowly, because the amount of heat would be reduced.

22. **(3)** *Application/Chemistry.* Cooling the beaker would slow down the rate of reaction and probably stop the product from splattering. The other choices are incorrect because they all would tend to speed up the reaction.

23. **(5)** *Analysis/Chemistry.* Because digestion is a series of chemical reactions, it is likely that these processes would slow down greatly without enzymes. The other choices all describe situations in which body processes are speeded up, resulting in heightened sensitivity to stimuli, more rapid body functions, and the inability to sleep.

24. **(2)** *Application/Chemistry.* Because the density of a gas decreases with temperature, warm air will rise above cool air. In order to circulate warm and cold air throughout a room, heated air should enter the room as low as possible and cooled air should enter the room as high as possible. Choice (1) is incorrect because it describes the reverse of the correct answer. Choice (3) would leave the upper half of the room without effective air conditioning and the lower half without effective heating. Choices (4) and (5) are irrelevant because they do not relate to the rising and sinking of the air.

Physics

1. **(4)** *Comprehension/Physics.* Throughout the 10-second interval, object D continues to move at a velocity of 10 meters/second. Thus its velocity is constant. If the object were standing still (1), it would have a velocity of zero. If it were coming to a stop (2), the line on the graph would slope downward toward zero. If it were accelerating (3), the line on the graph would slant upward. Choice (5) is irrelevant because the graph does not show direction.

2. **(3)** *Analysis/Physics.* According to the graph, object A began at rest (zero velocity) and now is increasing in velocity; object C is decreasing in velocity and eventually coming to a stop. Both of these conditions involve a change in motion that requires an outside force. During the first 4 seconds, both B and D are moving at constant velocity, a condition that, according to Newton's law, requires no action by a force.

3. **(5)** *Comprehension/Physics.* The slopes of both lines are parallel, meaning that velocity is increasing at the same rate for both objects. Choice (1) is irrelevant because the graph does not show position; choice (4) is irrelevant because the graph does not show direction. Choice (2) is incorrect because both objects are accelerating. Choice (3) is incorrect because object A is moving consistently at a greater velocity than object B.

4. **(1)** *Application/Physics.* Feathers obviously are much less dense than lead; however, a kilogram is a kilogram—thus the two masses would be the same. Because feathers have the lower density, the volume occupied by a kilogram of feathers would be much greater than the volume occupied by a kilogram of lead. Choice (2) is incorrect because objects with the same mass also would have the same weight. Choices (3) and (4) are incorrect because both the feathers and the lead have the same mass. Choice (5) is incorrect because the feathers and the lead do not have the same volume, but both do have the same weight.

5. **(1)** *Comprehension/Physics.* Choice (1) is correct because the passage implies that common measurements of movement—such as the speed of a car or airplane—are made with the assumption that the surface of the Earth is stationary. Choices (2), (3), and (4) are false statements. Choice (5) is not supported by the passage.

6. **(1)** *Application/Physics.* Because the rotation of the Earth is in the opposite direction to the motion of the car, the speed of the Earth must be subtracted. Choice (2) is incorrect because one must take into account the rotation of the Earth. Choice (3) would be correct if the car were moving in the same direction as the Earth, which is from west to east. Choices (4) and (5) are irrelevant.

7. **(4)** *Analysis/Physics.* A frame of reference is assumed to be fixed in place. If the frame of reference is moving, an object at rest will appear to be moving instead. Choices (1) and (2) are plausible, but they do not relate to the passage. Choice (3) is irrelevant, and choice (5) makes no sense.

8. **(2)** *Application/Physics.* Because sound waves travel considerably slower through air (a gas) than through a solid wall, putting several feet of space between the TV and the wall should help. Choices (1), (4), and (5) might work, but they do not relate to the information given in the passage. Choice (3) is incorrect because it will aggravate the problem.

9. **(5)** *Analysis/Physics.* The speed of light is 300,000 km/sec, whereas the speed of sound is only 346 meters/sec. Choices (1) and (3) are correct statements, but they do not relate to the question. Choices (2) and (4) are false statements.

10. **(2)** *Analysis/Physics.* Because sound waves cannot travel through a vacuum, music from a radio could not be heard on the moon. Choices (1) and (4) are not supported by the passage. Choices (3) and (5) are irrelevant.

11. **(3)** *Comprehension/Physics.* In order for work to be done, a force must cause an object to move. No motion is involved in standing still, holding a bag of groceries. All other choices involve some type of motion, even if the amount of movement is small.

12. **(2)** *Evaluation/Physics.* The passage states that the law of conservation of energy was changed once scientists realized that energy could be created from matter during nuclear reactions.

13. **(1)** *Analysis/Physics.* As the water rushes down the falls, friction causes heat energy, a type of kinetic energy. Choices (2) and (5) could be true, but they do not relate directly to the passage. Choices (3) and (4) have no basis in fact.

14. **(1)** *Analysis/Physics.* Because the 12-lb. ball has more mass, it would take more force to move it—which means that more work would have to be done on the ball. The other choices might be true, but they do not relate to the information given.

15. **(5)** *Evaluation/Physics.* It is clear from the diagram that the angles of refraction are the same when the light enters mediums A and C. Choice (1) is a true statement, but it is not sufficient evidence to conclude that mediums A and C are the same. Choices (2) and (3) are irrelevant. Choice (4) is incorrect because the light passes from A to B to C, not from A to C.

16. **(3)** *Application/Physics.* Because light is bent as it passes from air into water, the silver dollar would not be exactly where it appears to be when viewed by a person looking straight down into the pond. The other choices are incorrect because the information and diagram do not support the idea of refraction changing the apparent size or color of an object or its vertical position.

17. **(5)** *Comprehension/Physics.* Because force is directly proportional to mass and acceleration, and because the acceleration of all three objects is the same, the force required to move the objects must be in the same ratio as the objects' masses: 4:2:1.

18. **(4)** *Comprehension/Physics.* The passage states that radiation damages and destroys cells, thus it is unlikely that cells—especially life-sustaining cells—would increase. All other choices refer to cell damage or loss or the death of an organism.

19. **(2)** *Comprehension/Physics.* Choice (2) is correct because the electric current produced in a Geiger counter in the presence of a radioactive substance can help indicate its location. Choice (1) is incorrect because an electric current is caused in the Geiger counter by the radioactivity; the Geiger counter does not measure electricity. A Geiger counter is not capable of the functions listed in the other choices.

20. **(5)** *Analysis/Physics.* According to the passage, the main reason for detecting and measuring radiation is to protect living things from its damaging effects. All the other choices are plausible, but they do not relate to the central idea of the passage.

21. **(4)** *Application/Physics.* Increasing the mass of the water B would enable the water to absorb more heat. Changing the iron parts to aluminum C would mean that the metal parts of the machine could absorb more heat without changing temperature, because aluminum has a higher specific heat than iron. Changing the mass of the steel bearings A would be useless because the steel bearings are not overheating.

22. **(4)** *Analysis/Physics.* Because water has a much higher specific heat than air, and because the mass of the water is great, it is logical that the water would change temperature much less rapidly than the surrounding air. All the other choices are plausible, but they do not relate to the information given.

23. **(1)** *Comprehension/Physics.* In the diagram, the objects of uniform shape all are balanced at their centers, whereas the baseball bat and mobile are not balanced at their centers. Choices (2), (3), and (5) contradict the correct answer. Choice (4) is irrelevant because the question deals with the shape, not the size, of an object.

24. **(3)** *Analysis/Physics.* The information states that water expands when it freezes. Choice (3) is the only one that uses this information to explain the cracked pots.

HALF-LENGTH PRACTICE TEST

1. **(3)** *Application/Physics.* A solar energy-collecting panel should absorb energy, thus a dark color, such as black, would work best.

2. **(4)** *Analysis/Earth Science.* Because warmer air can hold more water, the warmest air, such as that found at noontime, would have the capacity to hold the most water. This would lower the relative humidity accordingly, because the maximum amount of water the air <u>could</u> hold would increase, even though the amount of water actually in the air stays the same.

3. **(1)** *Application/Biology.* Both the bacteria and the pea plant benefit from the relationship.

4. **(2)** *Application/Biology.* The remora benefits by receiving food and protection from its relationship to the shark; the shark is neither harmed nor helped by the relationship.

5. **(4)** *Application/Biology.* The steer is killed, its food energy benefiting the person who eats the beef.

6. **(3)** *Application/Biology.* The mistletoe lives off the pine tree to the mistletoe's benefit and the pine tree's detriment, because the mistletoe uses nutrients the pine tree would have used.

7. **(5)** *Application/Biology.* The bread mold feeds on the nonliving bread to its benefit.

8. **(3)** *Application/Physics.* The only choice that would prove conclusively that the element is not a metal is (3). Choices (1), (2), and (4) are characteristics of metals. Choice (5) does not give a clue as to whether or not the element is a metal.

9. **(3)** *Comprehension/Biology.* A lion and a tiger cannot interbreed, or produce offspring together, because they are not of the same species.

10. **(1)** *Comprehension/Earth Science.* Lava originates in the lowest layer of the sketch, in the mantle.

11. **(2)** *Comprehension/Earth Science.* There are shells in the shallow sea bed, whereas there are none in the deep sea bed, indicating this is a difference between the two.

12. **(4)** *Analysis/Earth Science.* For calcite to form into limestone, it must drop out of warm water. This indicates the area was part of a warm region, such as that in the tropics, at the time the limestone was formed. This eliminates choices (1), (2), and (3). Choice (5) is irrelevant because the question has nothing to do with the formation of sandstone.

13. **(1)** *Evaluation/Earth Science.* As depicted in the sketch, the layers are straight across, showing none of the folds, bends, and displacement usually associated with an earthquake. The other choices tend to support the conclusion that the area has undergone an earthquake.

14. **(3)** *Comprehension/Biology.* A frog is an amphibian, which makes it a cold-blooded animal that takes on the temperature of its surroundings; thus its body temperature would be 60° F.

15. **(3)** *Analysis/Biology.* The graph shows a steady increase in population after the advent of agriculture and the rapid increase in population after the beginning of the industrial revolution. The other conclusions are false, according to the graph.

16. **(2)** *Comprehension/Physics.* Pitch and frequency have nothing to do with the loudness of a sound; only decreasing the amplitude would decrease the volume.

17. **(2)** *Analysis/Physics.* As the ambulance approaches, you hear more and more wave compressions per second, increasing the siren's frequency and thus raising its pitch. Choice (1) is unlikely. Choices (3), (4), and (5) have to do with amplitude and loudness, which have no effect on pitch.

18. **(4)** *Analysis/Physics.* The amplitude of a sound is determined by the strength of vibrations. The louder the sound, the greater the amplitude, and thus, the vibrations. Applying this information to a vibrating eardrum leads to the conclusion that the louder the sound, the harder the eardrum vibrates.

19. **(5)** *Evaluation/Physics.* The fact that sound waves can break or destroy matter, such as windows or kidney stones, is evidence that they are a form of energy. A vibrating guitar string is not in itself evidence, because the string is not a sound wave itself, although it creates sound waves in the air.

20. **(2)** *Analysis/Chemistry.* A catalyst reduces the energy required for a reaction. The body's use of catalysts during digestion helps it obtain energy more efficiently. You can assume that body temperatures remain constant throughout the reaction, eliminating choices (1) and (3). Simpler sugars are formed from carbohydrates, not proteins, eliminating choice (4). Digestion would be too slow at body temperatures without catalysts but it still would occur, eliminating choice (5).

21. **(2)** *Analysis/Biology.* DDT removes some members of a population, leaving those resistant members to reproduce; their offspring inherit their resistance so that the DDT becomes less and less effective over time.

22. **(4)** *Analysis/Biology.* By increasing the chances of survival for birds that might be their nieces or nephews, even the adults that have no young might leave genes in the gene pool.

23. **(1)** *Analysis/Biology.* The males' elaborate songs signal a likely genetic advantage, and the females' responsive behavior is the outcome of evolution.

24. **(5)** *Analysis/Biology.* Moving away from the light has survival value for the larva, because this response would tend to lead it into the fruit, but whether it moves up or down is unimportant, given its feeding habits.

25. **(4)** *Analysis/Biology.* Of the possible responses, only shifting its activity to the daytime would increase the prey's chances of survival, because it then could hide at night when the predator is active.

26. **(3)** *Analysis/Biology.* Feeding every other day would strike a balance between the prey's preference and its increased susceptibility to attack when feeding. Choices (1) and (5) would lead to starvation. Choices (2) and (4) would increase the number of animals killed by the predator.

27. **(4)** *Evaluation/Earth Science.* According to the graph, the peak years of major Western droughts coincide with every other sunspot minimum as far back as 1700. This eliminates choices (1), (2), and (3). Choice (5) is false according to the graph, because the droughts occur about every 20 years.

28. **(4)** *Application/Physics.* Because a strawberry is red, it absorbs all the wavelengths of visible light except for the color we see—red—which is reflected back to us.

29. **(5)** *Application/Chemistry.* Cured ham is a cured meat, which contains sodium chloride (salt) as a preservative.

30. **(1)** *Application/Chemistry.* Orange marmalade is a jam, which contains sugar as a preservative.

31. **(3)** *Analysis/Biology.* Hard-to-reach plaque and tartar can be removed only by a dentist. Even people who practice careful oral hygiene need to visit a dentist regularly to avoid gingivitis.

32. **(5)** *Evaluation/Biology.* Although choices (1) to (4) are facts, they do not demonstrate the seriousness of the situation. The fact that periodontitis is the major cause of tooth loss (5) shows how serious the dentists' concerns are.

33. **(3)** *Evaluation/Chemistry.* The fact that protein cannot be used as is by the body indicates that digestion involves a chemical change. The amino acids are able to be used by the body, a property protein does not share. Statement A is not evidence, because it shows only that protein and amino acids have something in common. Statement

B is not evidence, because it deals with the mechanics of digestion.

FULL-LENGTH PRACTICE TEST

1. **(5)** *Application/Physics.* The truck has the greatest mass of the choices and would be hardest to accelerate; thus it would require the most fuel to accelerate.
2. **(2)** *Comprehension/Earth Science.* Only volcanic rock is not the remains or imprint of a living thing, thus it is not a fossil.
3. **(2)** *Comprehension/Biology.* A skin cell takes 8 + 6 + 5 + 1 hours to complete its life cycle, which is 20 hours, or about one day.
4. **(4)** *Comprehension/Biology.* The cell duplicates its chromosomes in preparation for when it will split into two cells during mitosis. The duplication of chromosomes alone is not evidence enough to prove that it is either a cancer cell or involved in sexual reproduction, eliminating choices (1) and (3). The cell must be in its synthesis stage, during which it does not grow, eliminating choices (2) and (5).
5. **(1)** *Analysis/Biology.* The plant's stem tip is a growth tissue. Putting this together with the knowledge that a plant's stem grows upward leads to the conclusion that pinching the stem tip will slow upward growth.
6. **(1)** *Application/Biology.* Because the root tip is growing continually, it most likely would have cells exhibiting each stage of the cell life cycle. The other choices are examples of specialized cells that do not continually divide.
7. **(3)** *Analysis/Biology.* To keep the parts of the body functioning well, new cells must be able to take over the previous cells' functions.
8. **(2)** *Application/Chemistry.* The best coating for the drug would be one that would allow it to pass through the stomach, break down in the intestine, and then pass into the blood to the kidneys. It should not break down in the stomach, choice (1), nor be excreted before reaching the kidneys, choices (4) and (5). It should break down in the intestine, eliminating choice (3).
9. **(5)** *Comprehension/Biology.* The other choices indicate a person might be susceptible to heart attacks, not that a heart attack actually has occurred.
10. **(1)** *Comprehension/Earth Science.* According to the graph, chromium should last until around 2500, which is the longest of any of the metals listed.
11. **(2)** *Application/Earth Science.* According to the graph, copper is the most abundant nonferrous metal in the United States and thus would best suit the manufacturer's needs.
12. **(5)** *Analysis/Earth Science.* Although any of the other choices might be true, only this statement both fits the facts in the graph and addresses industry's concern at another oil shortage caused by foreign oil producers.
13. **(1)** *Evaluation/Earth Science.* Lead and aluminum have estimated lifetimes of less than 200 years and thus support the belief. Coal and chromium have estimated lifetimes of 500 years or more and thus refute the belief.
14. **(3)** *Comprehension/Biology.* The use of oxygen for cellular respiration is shown by a decreased amount of oxygen in the blood.
15. **(5)** *Application/Biology.* An injury to the pituitary gland reducing the production of TSH would cause less thyroxin to be available, because, according to the diagram, TSH is the hormone that stimulates the thyroid to produce thyroxin.
16. **(1)** *Application/Physics.* The diver feels increased pressure because the pressure exerted by fluids increases with the depth of the fluid.
17. **(5)** *Application/Physics.* Bernoulli's principle states that as the fluid speeds up, the pressure decreases. Because the air on top is flowing faster, the air pressure above decreases. The difference in pressure causes the airplane to lift.
18. **(2)** *Application/Physics.* According to Pascal's principle, by applying pressure to a piston in the master cylinder, the pressure in the brake fluid is increased. This added pressure is transmitted to every part of the fluid and its container, in this case the brake lines.
19. **(4)** *Application/Physics.* If you think of the river as a tube, the water flows swiftly where the tube's area is decreased, i.e., where the river is shallow. It flows slowly where the area is increased, which is the point where the river is deep. Thus the speed of the water depends on the river, or tube, area.
20. **(2)** *Application/Chemistry.* The more and longer the fruit comes in contact with the ethylene gas, the more damage will occur, in the form of rotting fruit. The best strategy would take this into account by keeping the apples away from any rotting fruit and removing damaged apples as often as possible.
21. **(2)** *Application/Biology.* Cold viruses can enter the body through the eyes and nose. There is no evidence that choice (1) causes colds. Choice (3) probably would lead to the virus being killed by stomach acid. A virus on the cheek would not enter the body unless the skin were broken, which eliminates choice (4). Simply working next to someone with a cold, choice (5), probably will not lead to a cold unless you get the virus on your hand and then rub your eyes or nose.
22. **(1)** *Comprehension/Biology.* The passage states that plants combine carbon dioxide and water in the presence of sunlight to produce oxygen and glucose.

23. **(5)** *Application/Biology.* The passage states that most leaves are flat to expose a greater area to sunlight. It is logical that a flat surface would work best for solar collectors as well.

24. **(5)** *Analysis/Biology.* Carbon dioxide enters the plant through the leaves. Because carbon dioxide is necessary in making glucose during photosynthesis, blocking the entry of carbon dioxide with an impenetrable coating would interfere with photosynthesis.

25. **(4)** *Evaluation/Biology.* This statement indicates that water cannot travel from one part of a moss to another. Therefore, it is logical to assume that each part of a moss must absorb water directly from the environment.

26. **(5)** *Analysis/Biology.* By giving the plants equal amounts of sunlight and water, the scientist eliminates the variables in choices (1) and (2). By choosing identical plants, the scientist eliminates the variable in choice (3). By choosing containers of equal size, the scientist eliminates the variable in choice (4). Choice (5) is the variable because availability of nutrients might differ between soil and water.

27. **(2)** *Evaluation/Earth Science.* It is an overgeneralization to say that the Northeast will be consistently drier than the Southeast; the diagram merely indicates that this will be true much of the time in the next 20 years.

28. **(3)** *Analysis/Physics.* The fact that a carefully placed razor blade floats yet a dropped razor sinks indicates that the impact is responsible for the sinking. It is logical to assume that the impact breaks forces between the molecules, because cohesion is what holds the water molecules together and allows the razor to float.

29. **(4)** *Application/Biology.* Excess amino acids are broken apart. One part is excreted in the urine and the other might be converted to fat, making A and B likely. Possibility C is unlikely, because the body breaks down any amino acids it doesn't use for synthesis. An oversupply would be broken into its parts, not used for increased growth and repair.

30. **(4)** *Evaluation/Biology.* Excess amino acids are broken apart into other components that are either used in other body processes or gotten rid of as waste. Therefore, they cannot be stored as is by the body. Choice (1) is untrue. Choices (2), (3), and (4) are statements of fact but are only part of the answer and are not, by themselves, sufficient evidence to support the conclusion.

31. **(5)** *Analysis/Biology.* In attempting to prove that life could have arisen from nonliving matter, the scientist had to guess what conditions were like on early Earth. He assumed that lightning existed and played a role in the beginning of life. Choices (1) and (2) are facts provable by chemical analysis, not assumptions. Choice (3) is a hypothesis that a further experiment could test. Choice (4) is irrelevant to the experiment; how often amino acids form from gases is not at issue.

32. **(3)** *Evaluation/Biology.* Because scientists know that amino acids form proteins, the next step would be to prove that proteins form single-celled organisms. The most conclusive proof for this belief would be to actually form such organisms from proteins.

33. **(3)** *Evaluation/Chemistry.* Chemical analysis is a tool that can be used to prove the content of a particular substance. It cannot prove statements that depend on personal judgment, such as statements C and D.

34. **(3)** *Application/Chemistry.* The smaller the particle size, the greater the surface area and the faster the material dissolves. Aspirin powder contains the smallest particles and thus would dissolve fastest.

35. **(2)** *Application/Earth Science.* The guidelines call for avoiding particles that could enter the line, eliminating choices (3), (4), and (5), which are all less than 2 millimeters in diameter. Large rocks also should be avoided, eliminating choice (1).

36. **(3)** *Application/Earth Science.* The particles should be small yet should be able to flow without sticking together, making sand the best choice.

37. **(2)** *Application/Earth Sciençe.* The particles should provide a firm surface and be nonsticky yet not too large, making pebbles the best choice.

38. **(4)** *Application/Earth Science.* The finest particle that can be seen by the unaided eye is silt.

39. **(4)** *Application/Earth Science.* The two finest particles are silt and clay. (Marine shale is in fact formed from silt and/or clay.)

40. **(5)** *Analysis/Biology.* The kind of nerve impulses transmitted appears similar for both weak and strong stimulus; the sketch shows a similar wave for both. The number of nerve impulses transmitted and the number of neurons transmitting differ between the two stimuli, thus B and C must be how the brain differentiates between a weak and a strong stimulus.

41. **(2)** *Analysis/Biology.* Based on the experience with the elm tree, the best preventive measure would be to plant a variety of trees, because the passage implies that it was the planting of many elms, close together, that allowed the disease to spread throughout the entire population. Epidemics are least likely if no single variety of any organism is found in one area.

42. **(1)** *Comprehension/Biology.* The diagram shows that oxygen, which is given off by plants, is used by animals, who in turn give off carbon dioxide that is used by plants. Choices (2), (3), and (4) are incorrect, as shown by the diagram. Choice (5) does not address the relationship between plants and animals.

43. **(4)** *Application/Biology.* The fish immediately would add to the level of carbon dioxide and lower the level of oxygen. Later, there might be an overgrowth of plants, leading first to increased oxygen production, then to decay and still more carbon dioxide production.

44. **(5)** *Analysis/Biology.* Burning fossil fuels increases production of carbon dioxide, and deforestation leads to decreased use of carbon dioxide and less production of oxygen by trees.

45. **(1)** *Analysis/Biology.* The diagram shows that burning fossil fuels consumes oxygen and produces carbon dioxide. The action would not adversely affect plant and animal life.

46. **(2)** *Evaluation/Biology.* The diagram shows the effects of burning fossil fuels; both plants and animals would suffer from the upset balance. Choices (1) and (3) are illogical. Choices (4) and (5) take into account only half of the cycle.

47. **(4)** *Application/Earth Science.* Because 760 is standard, the air pressure will be a little less at higher altitudes, such as a mountaintop.

48. **(5)** *Comprehension/Physics.* According to the diagram, nuclear energy is used to make heat energy, which is used to create steam energy, which is used to turn the steam turbine, using the energy of motion, which is used to make electricity. A nuclear reactor makes electricity; it does not use electricity.

49. **(4)** *Application/Biology.* The whales' digging prepares the sea floor for the amphipods, much as a plow prepares the soil for crops.

50. **(2)** *Analysis/Biology.* Because the whales dig up large portions of the sea floor, they might disturb an oil pipeline placed along the bottom.

51. **(1)** *Analysis/Biology.* Mining the Bering Sea floor for sand would disturb and probably kill large numbers of the amphipods on which the whales feed, leading to a decrease of the whale population as well.

52. **(1)** *Analysis/Biology.* The amphipods anchor the sand into place through the tubes they live in, and harming them would lead to increased erosion of the sand. Disturbing the ecosystem in the other ways listed also could affect the erosion rate, but not as quickly as by harming the amphipods.

53. **(3)** *Evaluation/Biology.* An environmental preservation group would not want to emphasize a negative aspect of allowing the walruses to feed undisturbed—that the walruses eat food that could be harvested by fishermen to benefit certain people.

54. **(3)** *Evaluation/Biology.* A virus can reproduce by using a cell to make other viruses. Reproduction is an attribute of living organisms. The other choices all argue against a virus being classified as a living organism.

55. **(2)** *Analysis/Earth Science.* Coal takes so long to form that it is nonrenewable as far as the present generation and many to come are concerned.

56. **(2)** *Comprehension/Chemistry.* The passage states that plastics have replaced glass in household products, implying that glass is not a petrochemical product.

57. **(5)** *Analysis/Chemistry.* A heat-resistant cooler would slow the conduction of heat from outside the cooler to food inside the cooler. There is no reason for the other items to be heat resistant.

58. **(4)** *Analysis/Chemistry.* The strength, lightness, flexibility, and durability of plastic would be important to choices (1), (2), (3), and (5). A table top, however, could be made of any number of materials that would be as good as or better than plastic.

59. **(3)** *Evaluation/Chemistry.* The fact that workers in certain petrochemical factories have an increased chance of illness supports the argument. The other choices all have a potential human health benefit and thus refute the argument.

60. **(1)** *Application/Biology.* Forest fires burn away old trees, making room for young seedlings. In a crowded, old forest, controlled fires are a safe way to facilitate this process. Choice (2) is false. Choice (3) supports the opposing view. Choice (4) is a side benefit, not a main reason. Choice (5) is not supported by the diagram.

61. **(5)** *Evaluation/Biology.* The information supports the generalization that heart attacks occur at certain times of day, eliminating choices (1), (2), and (3). The platelets tend to stick together after a long sleep, eliminating choice (4) and supporting choice (5).

62. **(2)** *Comprehension/Physics.* As the tubes become narrower, the water rises higher.

63. **(1)** *Analysis/Physics.* The sketch shows the water rising in the tubes. The transport of water from the roots to the leaves is the only instance given of water rising through a "tube" in the plant.

64. **(5)** *Analysis/Physics.* There is no evidence that water pressure is causing the water to rise in the tubes. If water pressure were causing the rise, the water in the tubes would be only as high as the water surface itself. Thus, placing more water in the container would have no effect on the amount of water in the tubes. (The water rises because of attraction between the water molecules and the glass molecules.)

65. **(2)** *Comprehension/Biology.* Calcium helps make the bones hard and heavy. A loss of calcium would make the bones softer and lighter.

66. **(2)** *Evaluation/Biology.* Because it is not known just how these factors affect IQ scores, comparisons based on them should not be made. There is no basis for any of the other statements.

SIMULATED TEST

1. **(5)** *Application/Physics.* Of the choices listed, early evening is the coolest time of day and thus the time when the least water would evaporate.

2. **(2)** *Comprehension/Earth Science.* The passage emphasizes the dangers of being outside in a blizzard. It seems logical that the worst thing to do in this situation would be to leave the car. All other choices would be recommended actions to take.

3. **(1)** *Application/Biology.* In discussing a lesson, a teacher is passing information on to the students.

4. **(4)** *Application/Biology.* The dog is exploring its new surroundings.

5. **(2)** *Application/Biology.* The chicken has learned to link pecking at the blue lever with receiving food.

6. **(3)** *Application/Biology.* The male bird is benefiting the female, their young, and itself by feeding her while she's nesting.

7. **(5)** *Application/Biology.* The child swings at the ball over and over until he learns to do the task correctly.

8. **(2)** *Application/Chemistry.* An increase in baking soda would lead to an increase of carbon dioxide and water, which in turn would cause the bread to rise higher, producing a light, airy bread.

9. **(3)** *Comprehension/Biology.* The paragraph states that the biologist chose these particular fish to mimic those workings of the ocean that keep it livable, or clean, healthy, and fertile.

10. **(5)** *Comprehension/Earth Science.* The arrow points to the upper part of the water saturation zone.

11. **(4)** *Application/Earth Science.* The top of a hill is the farthest place from the zone of saturation, which, if high enough, can cause water seepage into a basement.

12. **(4)** *Analysis/Earth Science.* Draining the marsh of water completely would lower the water table. Because the bottom of the well is above the bottom of the marsh, the water table would be below the well and the well would run dry. The tree's roots also would have to reach deeper to reach the water table. The house's basement would not become flooded because it already is above the water table.

13. **(1)** *Analysis/Earth Science.* Keeping both considerations in mind, the point to the left of the house is far enough from the well, yet close to the house, and thus a good place to put a septic tank.

14. **(1)** *Comprehension/Biology.* Drug prescription requires the knowledge of how drugs (chemicals) affect the body and its life (biological) processes.

15. **(1)** *Application/Biology.* If two normal people have children together, their children will be normal also, because neither one is carrying a gene for color blindness.

16. **(4)** *Comprehension/Physics.* The passage states that meat cooked in a microwave doesn't get hotter than the boiling point of water, a temperature too low to cause it to brown

17. **(5)** *Application/Physics.* Metal containers cannot be used in a microwave, eliminating choices (2) and (3). Sealed containers could cause explosions, eliminating choices (1) and (4) and leaving the china bowl as the correct answer.

18. **(2)** *Analysis/Physics.* An egg is a closed container. Steam could build up enough pressure in an egg to explode it. There is no evidence in the passage to support the other choices.

19. **(3)** *Evaluation/Physics.* When reheating food, you want to heat it evenly, a use for which a microwave is well-suited. The other choices all involve foods that brown during cooking, a use for which a microwave is unsuited. The foods would not appear as they usually do, and their flavors would be affected.

20. **(1)** *Application/Chemistry.* Ascorbic acid keeps produce looking fresh. Tossing the juice from a citrus fruit, the lemon, with the salad would spread the ascorbic acid around the best. Choice (2) would not be as effective. Choices (3), (4), and (5) are not good sources of ascorbic acid.

21. **(4)** *Application/Biology.* By applying knowledge of how organic material improves soil, the two homeowners can restore the quality of the soil on the path. Because the path formerly was heavily traveled, you can assume it is too compacted, indicating it would be improved by tilling to loosen the soil. Because organic matter improves soil by fertilizing it, adding fertilizer also would benefit the soil.

22. **(5)** *Comprehension/Biology.* Cross-breeding involves crossing different plants to produce a new variety. Only choice (5) fits this definition.

23. **(3)** *Comprehension/Biology.* Genetic engineering involves manipulating genes. Causing a cell to mutate is a type of genetic engineering.

24. **(1)** *Analysis/Biology.* The fact that the introduction of farm crops increases soil erosion implies that less soil erodes when the land is covered by natural vegetation. The best explanation for this is that the natural vegetation is better suited to holding the soil in place.

25. **(4)** *Analysis/Biology.* Choice (4) summarizes the advantages of genetic engineering given in the passage. The other choices are too limited in scope and deal with only one specific detail.

26. **(2)** *Evaluation/Biology.* The passage states that the increased need for pesticides to grow new strains of crops is an environmental hazard. Choices (1) and (4) are limitations, but they are not reasons to stop trying. Choice (3) is a value judgment. There is no basis for choosing (5).

27. **(5)** *Evaluation/Earth Science.* Site E is at a low altitude and near the ocean, both of which help keep the air relatively warm. Thus a garden placed at this location would avoid some of the frosts that hit gardens at the other sites.

28. **(1)** *Analysis/Physics.* Because friction between two surfaces—tire and road—is what slows the car down, suspending the car in air would greatly cut down friction. Statement B would add to the friction by increasing the surface area in contact with the road. Statement C is irrelevant to the question, because it is concerned with friction between the car's moving parts, rather than the tires and the road.

29. **(5)** *Application/Chemistry.* Because trypsin digests proteins in the small intestine and its rate of action is highest between 8 and 9, you can predict that the small intestine's pH is between 8 and 9.

30. **(3)** *Analysis/Chemistry.* Because trypsin is most effective in a slightly basic pH, it probably doesn't function in the stomach because conditions are not right for it; in other words, the stomach is too acidic for trypsin to function.

31. **(2)** *Analysis/Biology.* A valid experiment matches variables to ensure that no factor other than the one being tested affects the results.

32. **(4)** *Evaluation/Biology.* The results clearly show that coronary angioplasty aids in reducing factors that lead to heart attacks. From this the conclusion can be drawn that coronary angioplasty helps prevent heart attacks. The other choices might be true, but they do not follow from the experiment's results.

33. **(5)** *Evaluation/Chemistry.* Statements A and B could be true of a carbon-based plastic or metal. But the fact that the pipe feels hot is evidence that it can conduct heat—a property of a metal, not plastic.

34. **(1)** *Analysis/Chemistry.* The soup's temperature dropped in the refrigerator, causing some of the salt to come out of solution and form crystals. A chemical change would change the salt into a new substance, eliminating choice (2). The soup could not evaporate much overnight, eliminating choice (5).

35. **(5)** *Comprehension/Earth Science.* The main implication of the changes brought about by the greenhouse effect is that life on earth is threatened. The other choices are statements of fact.

36. **(2)** *Comprehension/Earth Science.* Cutting down and removing an acre of trees would cause less carbon to be removed from the air, because the trees' photosynthetic activities would be lost. Less would be returned, because the trees' cellular respiratory activities also would be lost.

37. **(1)** *Analysis/Earth Science.* Erosion breaks down materials, such as limestone, that contain carbon. Statement B is incorrect because ponds absorb very small amounts of carbon compared to the ocean, and new ones form as old ones fill in. Statement C is incorrect because an interference with plant growth would decrease atmospheric carbon, not increase it.

38. **(3)** *Analysis/Earth Science.* Increased plant growth would lead to larger crops, thereby increasing farm productivity. Choice (1) is incorrect because an increased water level would not necessarily lead to more fish and seafood. Choice (2) is incorrect because carbon dioxide decreases air quality for oxygen-breathers, such as ourselves. Choice (4) is incorrect because, although the polar ice caps would melt, the fresh water they contain would be diluted by salt water, leading to no net increase of fresh water. There is no basis for choosing choice (5).

39. **(2)** *Evaluation/Earth Science.* The greenhouse effect warms the earth. Only choice (2) indicates a warming trend in the past century.

40. **(4)** *Analysis/Biology.* Meiosis cuts by half the number of chromosomes contributed by the sperm and egg, so that the end result is 46 chromosomes. If this didn't occur, the end result would be 92 chromosomes, then 184, and so on for each generation.

41. **(5)** *Analysis/Biology.* Assuming no other variable, the difference between breast-feeding and bottle-feeding babies is the action of their mouths. If this is the only difference, it is logical to conclude that it contributes to the straighter teeth observed in breast-fed children.

42. **(1)** *Comprehension/Biology.* The picture shows that the amount of DDT in the body builds up as it moves up the food chain from the smaller organisms to the larger organisms. Choice (2) is false, and choices (3), (4), and (5) are specific facts that do not demonstrate the relationship.

43. **(4)** *Application/Biology.* By comparing the weights of a seagull and a person, you can predict that a person who often eats DDT-contaminated clams would show much less of a concentration than the smaller seagull, yet more than a single clam. A concentration of 20 × DDT found in smaller organisms fits this prediction.

44. **(2)** *Analysis/Biology.* According to the picture, DDT is passed on from organism to organism, indicating it does not break down easily. Thus it will be eliminated gradually from living organisms. The ban of DDT would have little effect on the insect populations, because other insecticides still might be used, eliminating choices (3) and (4). There is nothing in the picture to indicate that DDT kills clams, making it unlikely that a ban would result in an increase in clams, thus eliminating choice (5).

45. **(3)** *Evaluation/Biology.* The only positive quality mentioned is that DDT kills insects. The other choices are all negative aspects of the use of DDT and would support the argument.

46. **(5)** *Evaluation/Biology.* DDT is shown affecting all the organisms in the sea ecosystem. The picture does not show seagulls dying, eliminating choice

(1). Other organisms absorb DDT, eliminating choice (2). There is no evidence for choices (3) or (4).

47. **(1)** *Application/Earth Science.* The cooler morning air would cause water to precipitate out of the air onto the grass. Because it is summer, temperatures would not be cold enough to form frost, ice, or snow. Although rain is a possibility under such conditions, the water is more likely dew, because the sky is cloudless.

48. **(2)** *Comprehension/Physics.* The arrows show that a force downward on the smaller piston results in an upward force on the larger piston, which would cause the car to move upward as well.

49. **(2)** *Application/Biology.* The rabbit population increased rapidly for the same reason the lamprey's did—the rabbits had no natural enemies in their new environment.

50. **(3)** *Analysis/Biology.* If a few sea lampreys survive, they will pass on to their offspring the genetic traits that allowed them to survive, and succeeding generations of lampreys will be immune to the poison.

51. **(5)** *Analysis/Biology.* The barriers should allow water and small fish to pass through so as not to disrupt the local ecosystem more than is necessary. The other choices are disadvantages.

52. **(4)** *Analysis/Biology.* The passage states that fish die when the pesticide's concentrations increase, making statement A correct. Statement B is correct because the chemical takes a long time to break down in the environment. Statement C is unlikely because an increase in the pesticide's occurrence in the environment doesn't necessarily mean that less is required to kill the lampreys.

53. **(1)** *Evaluation/Biology.* The fact that other fish decreased as lampreys increased is circumstantial evidence that the two are related. Choices (2) and (3) are evidence that pollution and overfishing also were responsible for the fish decline. Choices (4) and (5) help explain the lamprey's rapid spread through the Great Lakes.

54. **(2)** *Evaluation/Biology.* By comparing athletes active in the same sport and showing a lower injury rate for the group that limbered up first, the argument can be supported. Choice (1) is irrelevant to the argument, as is choice (5). Choices (3) and (4) cannot be used because they compare athletes in different activities.

55. **(2)** *Analysis/Earth Science.* Because the first type of volcano is created through separation and the second when the Earth's surface is coming together, it seems logical that pressure differences influence the force of the eruptions.

56. **(2)** *Application/Chemistry.* In solution with water, salt separates into electrically charged particles.

57. **(1)** *Application/Chemistry.* The treatment reduces the fabric's ability to burn.

58. **(4)** *Application/Chemistry.* Water is fairly hard to break down into other substances, thus it is considered a fairly stable substance.

59. **(5)** *Application/Chemistry.* Sodium hydroxide reacts vigorously with water, a test substance.

60. **(3)** *Analysis/Biology.* Agricultural advances have increased the food supply. Medical advances have reduced the death rate. These factors together have led to the increased population growth.

61. **(3)** *Evaluation/Biology.* A dieter is most concerned about calories and less concerned about price and flavor, choices (1) and (2). The passage says nothing about lower grades of beef, choice (4). The passage does not address food preparation, choice (5).

62. **(1)** *Evaluation/Physics.* Only choice (1) is true of both circuits. Choice (2) is untrue, according to the diagram. Choices (3) and (4) are true only of the series circuit. Choice (5) is an unlikely occurrence.

63. **(2)** *Application/Physics.* The electrons in a series circuit have only one path to flow along. If the path is interrupted at any point, the entire series is disconnected. This would cause all the appliances on the circuit to go off.

64. **(4)** *Analysis/Physics.* Because homes are wired in parallel, an overload would need to be stopped near the power source. Otherwise, some of the electrons could keep flowing along alternate routes, which could lead to overheating of the wires involved.

65. **(5)** *Comprehension/Biology.* A person who works outdoors is exposed to more sun, one of the risk factors of skin cancer. A person with light skin also is at a higher risk of skin cancer. Of the five choices, the fisherman is at highest risk. The x-ray technician and the person who has x rays taken are not exposed to enough radiation to have a problem.

66. **(3)** *Application/Biology.* Various sizes, colors, and shapes would stimulate the eye more the way the real world does, which could help decrease the incidence of nearsightedness in people who read a great deal.